The Prepper's Canning & Preserving Guide

The Complete Guide to Long-Term Food Preservation, Easy Recipes for
Dehydration, Pickling, Water and Pressure Preserving.

Survive After the Collapse of Society

Gary Gordon

© Copyright 2024 Gary Gordon all rights reserved.

This document is geared towards providing exact and reliable information with regard to the topic and issue covered. The publication is sold with the idea that the publisher is not required to render accounting, officially permitted, or otherwise qualified services. If advice is necessary, legal or professional, a practiced individual in the profession should be ordered. From a Declaration of Principles which was accepted and approved equally by a Committee of the American Bar Association and a Committee of Publishers and Associations.

In no way is it legal to reproduce, duplicate, or transmit any part of this document in either electronic means or in printed format. Recording of this publication is strictly prohibited, and any storage of this document is not allowed unless with written permission from the publisher. All rights reserved.

The information provided herein is stated to be truthful and consistent, in that any liability, in terms of inattention or otherwise, by any usage or abuse of any policies, processes, or directions contained within is the solitary and utter responsibility of the recipient reader. Under no circumstances will any legal responsibility or blame be held against the publisher for any reparation, damages, or monetary loss due to the information herein, either directly or indirectly.

Table of content

Preface 6
   Introduction to canning and preserving 6
   Importance of self-sufficiency in emergency preparedness 8
   Overview of the contents 10

Chapter 1: Basics of Canning and Preserving 14
   Understanding canning and preserving 14
   Historical context and modern relevance 17
   Key benefits for preppers 18

Chapter 2: Essential Tools and Equipment 21
   List of necessary canning supplies 21
   Overview of preserving tools 23
   Maintenance and care of equipment 25

Chapter 3: Food Safety and Hygiene 28
   Principles of food safety in canning and preserving 28
   Recognizing and preventing spoilage and botulism 30
   Best practices for cleanliness 32

Chapter 4: The Science of Canning 35
   How canning preserves food 35
   Different methods: water bath canning vs. pressure canning 37
   The role of acidity and sugar in preservation 39

Chapter 5: Preparing Your Produce 42
   Selecting and preparing fruits and vegetables 42
   Pre-treatment methods to enhance preservation 44
   Tips for maintaining flavor and texture 46

Chapter 6: Water Bath Canning 49
   Step-by-step guide to water bath canning 49

    Recipes for jams, jellies, and marmalades 51

    Canning high-acid foods 54

Chapter 7: Pressure Canning 56

    Detailed instructions for pressure canning 56

    Safety considerations and troubleshooting 58

    Recipes for meats, vegetables, and low-acid foods 60

Chapter 8: Dehydration and Freezing Techniques 64

    Basics of dehydrating fruits, vegetables, and meats 64

    How to use a dehydrator effectively 66

    Freezing techniques and best practices 69

Chapter 9: Fermentation and Pickling 72

    Introduction to fermentation 72

    How to make pickles, sauerkraut, and other fermented foods 74

    Health benefits of fermented foods 77

Chapter 10: Storing and Organizing Preserved Foods 80

    Best practices for labeling and storage 80

    Creating an inventory system 82

    Tips for maximizing shelf life 84

Chapter 11: Advanced Preserving Techniques 87

    Advanced recipes and techniques for experienced preppers 87

    Using preserves in everyday meals 89

    Making the most of seasonal abundance 92

Chapter 12: Building a Prepper's Pantry 95

    Designing a storage space for preserved goods 95

    Stockpiling strategies for long-term preparedness 97

    Regular maintenance of the pantry 100

Conclusion 103

    Recap of the skills and knowledge gained... 103

    Encouragement to practice and experiment 105
    Final thoughts on the role of preserving in prepping 107
Appendices 109
    Troubleshooting guide 109
    Glossary of terms 112

Preface

Introduction to canning and preserving

Canning and preserving are age-old techniques that have been used for centuries to extend the shelf life of food, ensuring a reliable supply during times when fresh produce is not available. These methods not only provide practical benefits but also enhance food security, particularly important for individuals focused on emergency preparedness—commonly known as preppers.

The Importance of Canning and Preserving

In the modern context, canning and preserving have regained popularity not only among preppers but also among those interested in sustainability, reducing food waste, and taking control of their food sources. These techniques allow you to keep and enjoy seasonal fruits and vegetables all year round, capture the peak freshness and nutrition of produce, and create a stockpile of food that is invaluable in emergency situations.

For preppers, the ability to preserve food is a fundamental aspect of preparedness planning. It provides self-sufficiency, reduces dependency on commercial food supply chains, and can be a lifeline during extended power outages, natural disasters, or other disruptions.

Canning vs. Preserving

While often used interchangeably, "canning" and "preserving" encompass a variety of techniques:

- Canning involves processing food in closed glass jars to kill microorganisms that cause food to spoil. This method uses heat and vacuum seals to preserve food, typically through water bath canning for high-acid foods like fruits and tomatoes, or pressure canning for low-acid foods like meats and vegetables.
- Preserving, more broadly, includes not just canning but also drying, freezing, fermenting, and pickling. Each method has its unique benefits and is suitable for different types of food.

Benefits of Learning Canning and Preserving

1. **Nutritional Advantages:** Preserved foods retain most of their nutrients, making them a healthy option during off-seasons.

2. **Economic Savings:** By buying in bulk during peak season and preserving, you can save money and have a variety of ingredients available at your fingertips.
3. **Flavor and Quality Control:** You control the ingredients, avoiding the excessive salts, sugars, and preservatives often found in commercially preserved foods.
4. **Reduced Food Waste:** Canning and preserving allow you to use more of what you buy or grow, turning ripe fruits and vegetables that might otherwise spoil into long-lasting preserves.

Getting Started

Starting your journey into canning and preserving doesn't require extensive expertise or equipment; with a few basic tools and some simple techniques, you can begin making your own preserves right at home. This guide will walk you through everything from selecting the right tools and understanding the science of canning to mastering various preserving methods and creating a well-stocked prepper's pantry.

As we dive into the following chapters, you'll learn the detailed processes, safety guidelines, and best practices to ensure success in your canning and preserving endeavors. Whether you're a novice looking to try your hand at making your first jar of jam or a seasoned prepper eager to enhance your skills, this guide aims to equip you with the knowledge and confidence to sustainably manage your food supply.

Importance of self-sufficiency in emergency preparedness

Self-sufficiency is a cornerstone of emergency preparedness, equipping individuals and communities with the resources and skills necessary to withstand crises without relying on external support. In the realm of food security, the ability to source, prepare, and preserve your own food is invaluable. This section discusses why self-sufficiency is critical in emergency preparedness, especially focusing on the role of canning and preserving.

Reducing Reliance on External Systems

Modern life is heavily dependent on complex, interconnected supply chains that deliver food and other necessities. While efficient, these systems are also vulnerable to disruptions caused by natural disasters, economic downturns, or global events like pandemics. By mastering the art of canning and preserving, preppers can minimize their dependency on these fragile supply chains, ensuring access to food regardless of external circumstances.

Ensuring Food Availability During Disasters

During emergencies, conventional food sources may become unavailable. Stores can quickly run out of supplies, and if roads are impassable, resupply may not be an option. Having a pantry stocked with canned and preserved foods ensures that you have a diverse and nutritious food supply on hand when you need it most, providing peace of mind and the ability to focus on other aspects of crisis management.

Enhancing Nutritional Security

Canned and preserved foods maintain much of their nutritional value, making them an excellent resource during times when fresh produce is not available. By preserving a variety of fruits, vegetables, meats, and other foods, preppers can ensure a balanced diet, rich in essential vitamins and minerals, even during prolonged emergencies.

Economic Efficiency

Investing in canning and preserving can also be economically advantageous. By purchasing produce in bulk during peak seasons and preserving it, you can avoid the premium prices charged for out-of-season or scarce items. Additionally, preserving food helps reduce waste, as you can process and store items that might otherwise spoil if bought in bulk.

Building Community Resilience

Self-sufficiency in emergency preparedness isn't just about individual or family readiness; it extends to community resilience. Communities that emphasize skills like canning and preserving can better support each other during crises. Sharing resources, knowledge, and preserved goods can strengthen communal ties and collective survival strategies.

Empowering Personal and Family Preparedness

Learning and practicing canning and preserving empower individuals and families to take charge of their emergency preparedness. This empowerment fosters a proactive mindset that is critical in managing unexpected situations. It also encourages the development of other survival skills that complement food preservation, such as gardening, foraging, and water purification.

The importance of self-sufficiency in emergency preparedness cannot be overstated, particularly concerning food security. Canning and preserving are vital skills that enhance self-reliance, ensuring that regardless of what happens in the world outside, you and your loved ones can have access to nutritious, home-preserved food. As we delve deeper into the specifics of canning and preserving techniques in the following chapters, remember that each jar you seal is not just food saved—it's a step toward greater independence and security in uncertain times.

Overview of the contents

"The Prepper's Canning & Preserving Guide" is designed to equip readers with the knowledge and skills needed to effectively can and preserve food at home, enhancing their self-sufficiency and preparedness for emergencies. This book is structured to guide you step-by-step through the various aspects of canning and preserving, ensuring you gain both the confidence and competence to build your own pantry of preserved foods. Here is a brief overview of what each chapter covers:

Chapter 1: Basics of Canning and Preserving

- Introduces the fundamental concepts of canning and preserving, explaining the importance of these techniques in building a self-sufficient lifestyle.

Chapter 2: Essential Tools and Equipment

- Details the necessary equipment and tools required for canning and preserving, including how to select, use, and maintain them effectively.

Chapter 3: Food Safety and Hygiene

- Discusses the critical aspects of food safety, including how to prevent foodborne illnesses and ensure that preserved foods are safe to consume.

Chapter 4: The Science of Canning

- Explains the scientific principles behind canning, including the importance of temperature, acidity, and the sealing process in preserving food.

Chapter 5: Preparing Your Produce

- Covers the best practices for selecting and preparing produce for canning, aiming to retain the best flavor and nutritional value.

Chapter 6: Water Bath Canning

- Provides a comprehensive guide to water bath canning, ideal for high-acid foods, with step-by-step instructions and recipes.

Chapter 7: Pressure Canning

- Focuses on pressure canning, necessary for preserving low-acid foods, detailing the process and safety measures to follow.

Chapter 8: Dehydration and Freezing Techniques

- Explores alternative preservation methods such as dehydration and freezing, which can be used in conjunction with canning or independently.

Chapter 9: Fermentation and Pickling

- Introduces fermentation and pickling, traditional methods that not only preserve food but also enhance its nutritional value and flavor.

Chapter 10: Storing and Organizing Preserved Foods

- Offers strategies for effectively storing and organizing your canned goods to maximize shelf life and ease of use.

Chapter 11: Advanced Preserving Techniques

- Delves into more complex preserving techniques and recipes for those looking to expand their skills beyond basic canning.

Chapter 12: Building a Prepper's Pantry

- Guides readers in setting up a comprehensive prepper's pantry, focusing on long-term storage and accessibility.

Conclusion

- Summarizes the skills learned throughout the book and encourages readers to continue practicing and refining their canning and preserving abilities.

Appendices

- Includes additional resources such as troubleshooting tips, a glossary of terms, and recommended further reading to help readers deepen their understanding of canning and preserving.

This overview serves as a roadmap to navigate through the book, ensuring that each chapter builds upon the previous one to develop a thorough understanding of canning and preserving as vital components of emergency preparedness. Whether you are a novice looking to learn these skills or an experienced prepper aiming to enhance your food storage solutions, this guide provides all the information you need to be successful.

Chapter 1: Basics of Canning and Preserving

Understanding canning and preserving

Canning and preserving are traditional methods used to extend the shelf life of food products. These techniques are crucial for anyone looking to maintain a reliable food supply throughout the year, particularly those preparing for situations where fresh produce is scarce. This chapter introduces the fundamental concepts of canning and preserving, providing a solid foundation for understanding their importance and practical application.

What are Canning and Preserving?

Canning is a method that involves processing food in airtight containers, typically glass jars, to extend its shelf life by eliminating microorganisms that cause spoilage. This process involves heating jars of food to kill bacteria, yeasts, and molds, and then sealing them to create a vacuum that prevents other microorganisms from entering and spoiling the food.

Preserving encompasses a broader range of techniques beyond canning, including drying, freezing, fermenting, and pickling. Each method has different benefits and is suitable for different types of food. For example, drying removes moisture to inhibit the growth of microorganisms, while fermenting and pickling involve creating an environment that favors beneficial bacteria, enhancing both preservation and flavor.

Historical Context

The history of canning and preserving dates back centuries and has played a critical role in human survival and civilization. Early methods of preservation included smoking, salting, and fermenting. The modern process of canning was developed in the late 18th century by Nicolas Appert, a French chef and confectioner, as a response to a challenge from the French government to preserve food for its army and navy.

Benefits of Canning and Preserving

1. **Extended Shelf Life**: Both techniques significantly extend the shelf life of food, making it possible to enjoy seasonal fruits and vegetables throughout the year.
2. **Nutrition**: When done properly, canning and preserving can lock in vitamins and minerals, making these foods a valuable nutrient source during off-seasons or emergencies.
3. **Flavor**: Many people find that preserved foods, such as jams, pickles, and canned tomatoes, develop rich flavors that enhance a variety of dishes.
4. **Economy**: Canning and preserving allow you to buy foods in bulk when they are in season and less expensive, and then store them for use throughout the year.
5. **Self-Sufficiency**: These skills enable individuals to rely less on commercial food sources, which is especially valuable in times of supply chain disruptions or personal financial constraints.

How Canning and Preserving Work

Canning

- **Water Bath Canning**: Suitable for high-acid foods such as fruits and pickles. The jars are filled with food and a water-based liquid, sealed, and then boiled in a water bath to create a vacuum seal.
- **Pressure Canning**: Necessary for low-acid foods like vegetables and meats. This method uses a specialized pressure canner to achieve higher temperatures than boiling water alone, ensuring that all potentially harmful organisms are destroyed.

Preserving

- **Drying**: Food is dehydrated to remove moisture that bacteria, yeasts, and molds need to grow.
- **Freezing**: Slows down the activity of harmful bacteria and other microorganisms by freezing and storing food at below-zero temperatures.
- **Fermenting and Pickling**: Involves creating an acidic environment where harmful bacteria cannot thrive but beneficial bacteria can, enhancing both preservation and probiotic content.

Understanding the basics of canning and preserving is essential for anyone looking to enhance their food security and independence. These methods not only provide practical benefits but also contribute to healthier eating habits by allowing you to control the ingredients in your diet. As we explore more detailed techniques and safety guidelines in subsequent chapters, you'll be equipped with the knowledge to start your own canning and preserving journey, ensuring a steady supply of food regardless of external circumstances.

Historical context and modern relevance

Historical Context

Canning and preserving food are practices steeped in history, essential for survival and the advancement of civilizations. The roots of these methods trace back to ancient times when early humans sought ways to ensure a stable food supply across seasons and during travels.

Ancient Methods: Long before modern canning, ancient civilizations employed techniques like drying, smoking, fermenting, and salting to preserve meats, fruits, and vegetables. These methods were crucial for survival during winters, droughts, and long journeys.

Revolutionary Developments: The turning point in food preservation came in the early 19th century with the invention of canning by Nicolas Appert in France. Appert, responding to a challenge from the French government, developed a method to preserve food in sealed glass jars placed in boiling water. This innovation won him the government prize in 1810, fundamentally changing food storage.

Spread and Innovation: Appert's method quickly spread across Europe. Soon after, the English inventor Peter Durand introduced the use of tin cans, which were easier to transport and less fragile than glass containers. This innovation led to the development of the canning industry, further revolutionized during World War I and II, when the need for stable food supplies became critical.

Modern Relevance

In today's world, canning and preserving have transcended their origins as mere survival techniques, morphing into popular hobbies and key components of sustainable living practices.

Emergency Preparedness: The modern prepper movement emphasizes the importance of self-sufficiency in uncertain times. Canning and preserving are fundamental for building an emergency food supply, providing stability when store access is interrupted by natural disasters, pandemics, or economic instability.

Sustainable Living: With growing awareness of environmental issues and food waste, canning and preserving allow individuals to reduce their carbon footprint. By preserving seasonal produce, food waste is minimized, and the environmental impact of transporting out-of-season produce is reduced.

Health and Nutrition: Modern consumers are increasingly wary of the additives and preservatives in commercially processed foods. Canning and preserving at home lets individuals control the ingredients in their food, leading to healthier dietary choices free from unnecessary chemicals.

Cultural and Community Impact: Canning and preserving have seen a revival as part of the local food movement. Communities are coming together through shared canning workshops and swap meets, where techniques are passed down and culinary traditions preserved.

Economic Benefits: During economic downturns, canning and preserving provide a cost-effective way to maximize food budgets. By buying in bulk during peak seasons and preserving, households can enjoy high-quality food year-round without premium prices.

The historical context and modern relevance of canning and preserving highlight their pivotal role in both past and present societies. As we continue to face global challenges, these time-tested techniques remain as vital as ever, providing food security, promoting sustainability, and enriching our connection to our food and each other. In subsequent chapters, we will explore the practical application of these methods, ensuring readers are well-equipped to apply these historic practices to modern-day needs.

Key benefits for preppers

For preppers, the practice of canning and preserving food is not just a hobby but a strategic component of their emergency preparedness plans. The ability to store food long-term provides numerous benefits, ensuring that individuals and families can remain self-reliant in various scenarios, from economic downturns to natural disasters. This section outlines the key benefits of canning and preserving for preppers.

Extended Food Security

One of the primary advantages of canning and preserving is the significant extension of food's shelf life. Foods that might only last a few days or weeks in their natural state can be safely consumed for months or even years when properly canned or preserved. This long-term storage capability is crucial for preppers, who aim to maintain a stable food supply regardless of external circumstances.

Independence from Supply Chains

Preppers often seek to reduce their dependence on traditional food supply chains, which can be disrupted by various events, including severe weather, pandemics, or geopolitical tensions. By mastering canning and preserving techniques, preppers can create a buffer against these disruptions, ensuring access to essential nutrients without the need for frequent grocery store visits.

Cost-Effectiveness

Canning and preserving can also be economically beneficial, especially when utilizing seasonal produce or bulk purchasing. This approach allows preppers to buy food at lower prices during peak season and preserve it for use throughout the year, avoiding the cost of out-of-season or emergency purchasing, which often comes at a premium.

Nutritional Retention

Preserved foods maintain much of their nutritional value, making them an excellent resource during times when fresh produce is not available. By canning and preserving a variety of fruits, vegetables, meats, and other staples, preppers can ensure a well-rounded diet, rich in essential vitamins and minerals, even during extended periods of self-reliance.

Preparedness for Evacuation

Canned goods are portable, making them an ideal choice for emergency evacuation kits. Unlike frozen foods, canned items do not require electricity, and unlike fresh produce, they are not perishable over short periods. This portability ensures that preppers can maintain a ready supply of nutritious food even on the move.

Diverse Food Options

Preserving various types of food by canning, drying, and fermenting not only secures a stable food supply but also prevents menu fatigue during prolonged periods of reliance on stored food. Preppers can enjoy a diverse array of flavors and textures, making meals more enjoyable and varied, which is crucial for morale during challenging times.

Enhancing Self-Sufficiency Skills

The process of learning and practicing canning and preserving enhances overall self-sufficiency skills. It promotes a mindset of resourcefulness and preparedness that can be applied to other areas of life, such as gardening, water purification, and basic first aid.

The benefits of canning and preserving for preppers are clear and substantial. These techniques support a sustainable, independent, and proactive approach to emergency preparedness. As we explore these methods in more depth throughout this book, preppers will gain valuable insights and practical skills that enhance their readiness for any situation.

Chapter 2: Essential Tools and Equipment

List of necessary canning supplies

Canning is a rewarding process, but having the right tools and equipment is essential for safe and effective preservation. This chapter provides a comprehensive list of necessary canning supplies that will help both beginners and experienced canners ensure their efforts are successful and their food is safely preserved.

Basic Canning Supplies

1. **Canning Jars:** Glass jars specifically designed for canning are a must. They come in various sizes, typically ranging from half-pints to quarts, and are reusable if in good condition.
2. **Lids and Bands:** The lids are crucial for sealing the jars; however, they are only meant for single-use when it comes to canning. Bands hold the lids in place during the processing and can be reused.

3. **Canner:** Depending on what type of food you plan to can, you might need one or both types of canners:
   - **Water Bath Canner:** Used for high-acid foods like fruits, jams, jellies, and pickles.
   - **Pressure Canner:** Essential for canning low-acid foods such as meats, vegetables, and meals like soups and stews.
4. **Jar Lifter**: A jar lifter is designed to safely lift hot jars out of boiling water or from a pressure canner.
5. **Canning Funnel**: A wide-mouth funnel helps in transferring food into jars without spilling, keeping the jar rims clean for a good seal.
6. **Bubble Remover/Headspace Tool**: This tool is used to remove air bubbles before sealing the jars. It also helps measure the proper headspace, which is crucial for ensuring a good seal.
7. **Magnetic Lid Lifter**: This tool allows you to remove canning lids from hot water safely, helping to keep them sterile.

Additional Useful Items

8. **Kitchen Timer:** Precision in processing times is critical in canning, so a reliable timer is indispensable.
9. **Clean Cloths and Towels**: Having plenty of clean cloths on hand is necessary for wiping jar rims, handling hot jars, and cleaning up spills.
10. **Non-Reactive Cooking Utensils**: When preparing your preserves, use non-reactive utensils made of stainless steel, silicone, or wood. Avoid reactive metals like aluminum or copper with acidic foods.
11. **Large Pot**: If you're starting without a dedicated canner, a large, deep pot can be used for water bath canning, provided it's deep enough to allow water to cover the jars completely.
12. **Chopping Tools**: Sharp knives and cutting boards are essential for preparing ingredients. A food processor or blender can also be handy for sauces and purees.
13. **Labels and Permanent Marker**: Labeling jars with the contents and date of canning will help you keep track of your inventory and use older stocks first.
14. **Oven Mitts or Gloves**: Protect your hands from heat when handling hot equipment and jars.

Equipping yourself with these tools will prepare you to tackle any canning project confidently. The initial investment in quality supplies will pay off in the safety and quality of your preserved foods. In the next section, we will delve into how to use these supplies effectively, ensuring you maximize their benefits as you build your home canning and preserving capabilities.

Overview of preserving tools

Preserving food involves a variety of techniques beyond canning, such as drying, freezing, fermenting, and pickling. Each method requires specific tools to ensure the process is efficient, safe, and results in high-quality preserved foods. This section provides an overview of the essential tools used in these different preserving methods, helping you to understand their functions and how they contribute to successful food preservation.

Drying Tools

1. **Food Dehydrator:** A food dehydrator is an electric appliance that evenly circulates hot air to remove moisture from foods. It's ideal for drying fruits, vegetables, herbs, and meats to make jerky. Dehydrators often come with multiple trays, allowing you to dry large batches at once.
2. **Oven**: For those without a dehydrator, a conventional oven can be set at a low temperature to dry foods. It's less energy-efficient and requires more attention, but it's a good starting point.
3. **Air-Drying Racks or Nets**: These are used for air-drying herbs and some fruits in a well-ventilated, dry area. They help maintain good air circulation around the food, preventing mold.

Freezing Tools

4. **Freezer:** A high-quality freezer is crucial for preserving food through freezing. Chest or upright freezers are preferred for their space and efficiency.

5. **Vacuum Sealer**: This device removes air from plastic bags or containers before sealing them, minimizing freezer burn and extending the shelf life of frozen foods.
6. **Freezer-Safe Containers**: These include plastic containers, glass jars specifically designed to withstand low temperatures, and heavy-duty freezer bags.

Fermentation and Pickling Tools

7. **Fermentation Crocks**: Traditional stoneware crocks are used for large-scale fermentation of vegetables, such as for making sauerkraut or kimchi. They often come with weights to keep vegetables submerged under brine.
8. **Airlock Lids**: These lids fit onto standard mason jars and are designed to let gases escape during fermentation while keeping air out, which is crucial for creating an anaerobic environment.
9. **Glass Jars**: Widely used for smaller batches of pickles and fermented foods. They are affordable, readily available, and come in various sizes.
10. **pH Meters or Test Strips**: These tools are useful for monitoring the acidity levels in pickles and ferments to ensure they are safe for storage.

Additional Useful Accessories

11. **Paring and Chef's Knives**: Essential for preparing fruits, vegetables, and meats for any type of preservation.
12. **Cutting Boards**: Provide a safe and clean surface for slicing and chopping ingredients.
13. **Measuring Cups and Spoons**: Ensure that you use precise amounts of salts, sugars, and seasonings, which is critical for successful preservation.
14. **Colander**: Useful for washing fruits and vegetables before preserving them.
15. **Mixing Bowls**: Needed for mixing ingredients or brining solutions.

Whether you're dehydrating, freezing, fermenting, or pickling, having the right tools can significantly affect the ease and success of your preservation efforts. Each tool plays a unique role in the process, from preparing ingredients to executing specific preservation techniques. As you build your collection of preserving tools, consider the types of foods you plan to preserve most frequently and choose tools that best suit your needs and space constraints. In the upcoming chapters, we will explore how to use these tools effectively to maximize the quality and longevity of your preserved foods.

Maintenance and care of equipment

Proper maintenance and care of canning and preserving equipment are crucial for ensuring food safety, extending the lifespan of your tools, and maintaining their effectiveness. This section will provide detailed guidance on how to take care of your canning and preserving equipment, covering everything from basic cleaning to storage and routine maintenance.

General Maintenance Tips

1. **Cleaning After Each Use:** Always clean your equipment immediately after use. Residual food particles can harbor bacteria and attract pests, and certain acidic foods can corrode or damage materials if left in contact for too long.
2. **Use Appropriate Cleaning Agents**: Use mild detergents and avoid abrasive scrubbers or harsh chemicals that can damage surfaces, especially on sensitive materials like rubber gaskets and plastic parts.
3. **Dry Thoroughly**: After washing, dry your equipment thoroughly to prevent rust and mold growth. This is particularly important for metal items like canning lids, jar bands, and tools used in fermentation.
4. **Inspect Regularly**: Before each use, inspect your equipment for signs of wear and tear. Look for cracks in jars, rust on metal parts, degradation of rubber seals, and other damage that could compromise the safety or effectiveness of your canning and preserving efforts.

Specific Equipment Care

Canning Jars and Lids

- Wash jars in hot, soapy water or in a dishwasher. Check each jar for nicks, cracks, or rough edges that could prevent sealing or cause breakage.

- Replace canning lids each time you use them for preserving. Bands can be reused but should be replaced if they show any signs of rust or deformation.

Pressure Canners and Water Bath Canners

- Check the rim and the inside of your canner for mineral deposits and rust. Clean with vinegar or a manufacturer-recommended cleaning agent.
- For pressure canners, inspect the vent, safety valve, and gasket annually. Replace the gasket as recommended by the manufacturer or if it shows signs of wear.
- Store the lid off the canner to allow air circulation and prevent odors.

Dehydrators

- Clean trays and screens after each use with warm, soapy water. Avoid using abrasive cleaners or pads.
- Inspect the fan and heating element regularly for dust and debris, which can reduce efficiency.

Vacuum Sealers

- Wipe the exterior and the sealing strip after each use. Use a damp cloth; avoid abrasive materials.
- Check the sealing strip for wear and ensure it's not compressed or damaged.

Fermentation Crocks and Airlock Lids

- Wash crocks and lids thoroughly with warm, soapy water. Rinse well to remove all soap residues, which can affect the fermentation process.
- Store crocks in a cool, dry place to avoid mold growth.

Freezers

- Defrost and clean your freezer at least once a year to maintain efficiency and prevent ice build-up.
- Check and clean the seal regularly to ensure it is tight and clean, which helps prevent energy loss.

Storage Tips

- Store all tools and equipment in a clean, dry place to prevent dust and moisture buildup.
- Hang large utensils such as ladles and jar lifters to avoid damage and save space.
- Keep sharp tools like knives and peelers in protective sheaths or a dedicated storage block.

Taking proper care of your canning and preserving equipment not only ensures the safety and quality of your preserved foods but also saves you money in the long run by extending the life of your tools. Regular maintenance, combined with careful cleaning and storage, will keep your equipment in top condition, ready for each canning and preserving season.

Chapter 3: Food Safety and Hygiene

Principles of food safety in canning and preserving

When it comes to canning and preserving food, safety is paramount. The process of storing food for long periods can pose risks if not done correctly, primarily due to the potential growth of bacteria, molds, and yeasts that can cause foodborne illnesses. This chapter outlines the essential principles of food safety to follow when canning and preserving, ensuring that your home-preserved foods are both delicious and safe to consume.

Understanding the Risks

1. **Botulism:** This is one of the most serious risks associated with improperly canned foods. Botulism is caused by a toxin produced by Clostridium botulinum bacteria, which can thrive in low-oxygen environments like sealed jars.

2. **Mold and Yeast Growth**: Improper sealing or storage can allow molds and yeasts to grow, spoiling the food and potentially causing illness.
3. **Chemical Contamination**: Using inappropriate materials in canning can lead to chemical contamination. For instance, using non-food-grade metals can result in the leaching of harmful substances into the food.

Key Food Safety Principles

1. Sterilization of Canning Jars and Lids

- Before use, all canning jars and lids must be sterilized to kill any existing pathogens. This is typically done by boiling them in water for at least 10 minutes.

2. Use of Proper Canning Methods

- High-Acid Foods: Foods like fruits and tomatoes can be safely canned using a water bath method, which involves boiling the jars in water.
- Low-Acid Foods: Vegetables, meats, and seafood must be canned using a pressure canner, which reaches higher temperatures than boiling water, ensuring all bacteria are destroyed.

3. Correct Processing Times and Temperatures

- Follow tested recipes and guidelines for the correct processing times and temperatures to ensure all pathogens are destroyed. Underprocessing can leave bacteria alive, while overprocessing can compromise the nutritional quality and taste of the food.

4. Maintaining Cleanliness

- Keep all work surfaces, utensils, and your hands clean when preparing and canning food to prevent the introduction of bacteria into your canned goods.

5. Avoiding Cross-Contamination

- Use separate cutting boards and utensils for different types of foods (e.g., raw meats and vegetables) to prevent cross-contamination.

6. Proper Sealing and Storage

- Ensure that jars are sealed properly to create an anaerobic environment that inhibits the growth of harmful bacteria. Store preserved foods in a cool, dark place to maintain their quality and safety.

7. Regular Inspection of Stored Foods

- Regularly check your preserved foods for any signs of spoilage, such as off smells, visible mold, or bulging lids, which can indicate bacterial growth.

8. Acidification

- Adding lemon juice or vinegar to low-acid foods can help raise their acidity, making them safer for water bath canning.

Adhering to these food safety principles is critical for anyone involved in canning and preserving. By understanding the risks and applying these guidelines diligently, you can ensure that your home-canned foods are not only nutritious and flavorful but also safe to eat. As we continue to explore specific canning and preserving techniques in the following chapters, keeping these safety practices in mind will be essential for successful and safe home preservation efforts.

Recognizing and preventing spoilage and botulism

Canning and preserving are effective methods for extending the shelf life of food, but they require careful handling to prevent spoilage and the dangerous risk of botulism. Understanding how to recognize signs of spoilage and implementing strategies to prevent botulism are critical aspects of safe food preservation practices. This section provides detailed guidance on these topics.

Recognizing Spoilage in Preserved Foods

Spoilage in canned and preserved foods can present in various ways. Here are some signs to watch for:

1. **Off Odors:** A foul or unusual smell when opening a jar is a clear indication of spoilage.
2. **Visible Mold or Yeast**: Any signs of mold or yeast growth inside the jar or on the food surface suggest contamination and spoilage.
3. **Gas Bubbles:** Bubbles moving in the jar or a fizzy sound upon opening can indicate fermentation or bacterial activity.
4. **Discoloration:** Any significant changes in the color of the food that were not present immediately after canning could be a sign of spoilage.
5. **Bulging Lids**: Lids that bulge or are domed outward suggest gas production inside the jar, a common sign of bacterial activity.
6. **Leakage**: Seeping liquid or stains on the outside of the jar can be caused by bacterial gases pushing liquid out.

Preventing Spoilage

To minimize the risk of spoilage, follow these best practices:

1. **Sterilize All Equipment:** Before canning, ensure all jars, lids, and tools are properly sterilized to eliminate existing microbes.
2. **Use Fresh, High-Quality Ingredients**: Spoilage is more likely if the food was near spoiling before it was canned. Always use fresh and undamaged ingredients for canning.
3. **Adhere to Recipes**: Use tried and tested recipes and follow them exactly, especially the recommended headspace, processing time, and temperature.
4. **Proper Sealing**: Ensure jars seal correctly and test seals when the jars have cooled. Improper sealing is a common cause of spoilage.

Recognizing and Preventing Botulism

Botulism is a rare but serious illness caused by a toxin produced by Clostridium botulinum bacteria. It can be fatal if not treated promptly.

Recognition:

- Home-canned foods are the most common source. The toxin is odorless and invisible, so detection based on visual inspection is not possible.
- Symptoms of botulism include blurred vision, slurred speech, difficulty swallowing, muscle weakness, and paralysis.

Prevention:

1. **Proper Canning Method**: Use a pressure canner for all low-acid foods, such as vegetables, meats, and seafood, to achieve the high temperatures needed to kill botulinum spores.
2. **Acidification**: For borderline low-acid foods like tomatoes, add acid in the form of lemon juice or vinegar to raise the acidity to a safer level.
3. **Avoid Inadequate Seals**: Ensure that all jars are sealed properly. Any unsealed jars should be refrigerated and used quickly or frozen.
4. **Store Correctly**: Keep canned foods in a cool, dark place and use them within a recommended period, typically one year for best quality.
5. **Boil Home-Canned Foods**: Before eating, boil home-canned foods for 10 minutes to ensure any present botulinum toxin is inactivated.

Recognizing the signs of spoilage and understanding the risks and prevention methods for botulism are essential for safely canning and preserving food at home. By following strict hygiene and processing guidelines, you can ensure that your preserved foods remain safe and delicious. As you gain more experience, maintaining these standards will become a routine part of your preserving practices, providing you and your family with a reliable and safe food supply.

Best practices for cleanliness

Maintaining a high level of cleanliness is crucial in the process of canning and preserving foods. Proper sanitation prevents the introduction and proliferation of harmful bacteria, yeasts, and molds that can cause food spoilage and foodborne illnesses. This section outlines the best practices for cleanliness that should be followed to ensure safe and effective canning and preserving.

Cleanliness in the Workspace

1. **Sanitize Work Surfaces:** Before starting the canning process, thoroughly clean and sanitize all work surfaces. Use a solution of bleach and water (1 tablespoon of unscented bleach per gallon of water) or a food-safe sanitizer. Wipe surfaces down and allow them to air dry to ensure a sanitary environment.
2. **Wash Your Hands Frequently:** Handwashing is one of the most effective ways to prevent the spread of bacteria. Wash hands with warm water and soap for at least 20 seconds before handling any food or canning materials, and repeat frequently, especially after touching anything that could contaminate your clean workspace.
3. **Use Clean Towels and Cloths:** Always start with fresh, clean towels and dishcloths. Using dirty or used cloths can reintroduce bacteria to clean surfaces and equipment. Consider using paper towels for drying hands and surfaces during the canning process to prevent cross-contamination.

Cleanliness with Equipment and Tools

4. **Sterilize Canning Jars and Lids:** Boil canning jars, lids, and bands for at least 10 minutes before using them to ensure they are free from bacteria. Keep them in hot water until they are used to prevent recontamination.
5. **Keep Utensils Sanitized:** Submerge spoons, ladles, funnels, and other utensils in boiling water for a few minutes before using them. This practice ensures that these tools are sterile when they come into contact with your food and canning jars.
6. **Regularly Clean and Inspect Canning Equipment:** Clean all parts of your canning equipment, such as jar lifters, pressure canners, and water bath canners, after each use. Inspect them for any signs of wear or rust that could harbor bacteria.

Preventing Contamination

7. **Avoid Double Dipping:** Once a utensil has been used for filling jars, do not allow it to come into contact with any other food or surfaces before it has been cleaned again. This practice helps to prevent the spread of bacteria from one batch to another.
8. **Manage Food Waste Promptly:** Dispose of food waste and trimmings immediately to avoid attracting pests or developing odors that can lead to contamination. Use a designated bin that is emptied and cleaned regularly.
9. **Control Traffic in the Kitchen:** During the canning process, limit the number of people who come in and out of the kitchen. This reduces the introduction of new contaminants into the area.

After the Canning Process

10. **Store Equipment Properly:** After cleaning, ensure all equipment is dry before storing it in a clean, dry area. Moisture can promote the growth of mold and bacteria.
11. **Label and Date Everything:** Use labels to mark the contents and date of canning on each jar. This not only helps in organizing but also ensures that older batches are used first, reducing the risk of using spoiled food.

Following these best practices for cleanliness will significantly enhance the safety and success of your canning and preserving efforts. By keeping your environment, tools, and techniques clean, you minimize the risk of contamination, ensuring that your preserved foods remain safe and high-quality for extended periods. This diligence in cleanliness is a critical component of effective food preservation.

Chapter 4: The Science of Canning

How canning preserves food

Canning is a time-honored method that extends the shelf life of food by using heat to kill potentially harmful microorganisms and an airtight seal to prevent new ones from entering and spoiling the food. This chapter explains the science behind canning, detailing how it effectively preserves food and ensures safety and longevity.

The Role of Heat

1. **Killing Microorganisms:** The primary objective of the heat process in canning is to destroy bacteria, yeasts, molds, and enzymes that cause food spoilage and potentially dangerous foodborne illnesses. The temperature reached during the canning process, especially in pressure canning, is sufficient to kill even the most resistant spores, like those of Clostridium botulinum, which causes botulism.

2. **Heat as a Preservation Tool:** Beyond killing microorganisms, heat also works to break down some of the natural compounds in food that can lead to spoilage. This alteration helps in stabilizing the food's color, flavor, and texture during storage.

The Importance of the Vacuum Seal

3. **Removing Air:** During the canning process, as jars are heated, the contents expand, and air is forced out. When the jars cool, a vacuum seal forms, significantly reducing the presence of oxygen inside the jar. Oxygen is necessary for many bacteria and fungi to grow; its absence helps preserve the food.
4. **Maintaining the Seal:** A secure vacuum seal prevents the entry of new air and contaminants into the jar. This seal is crucial for maintaining the sterility of the food product after canning. Any failure in the seal can allow bacteria to enter and spoil the food.

The Role of Acidity in Canning

5. **High-Acid vs. Low-Acid Foods:** The acidity of the food being canned plays a crucial role in determining the method of canning that should be used. High-acid foods (with a pH of 4.6 or lower), such as most fruits and pickles, can be processed using a water bath canner. The natural acids in these foods inhibit the growth of many bacteria.
6. **Pressure Canning for Low-Acid Foods:** Low-acid foods, such as meats and most vegetables, have a pH higher than 4.6 and do not naturally inhibit bacterial growth. Pressure canning is required for these foods because it reaches higher temperatures than boiling water, ensuring all bacteria, particularly botulinum spores, are destroyed.

Synergistic Effects of Canning Components

7. **Interaction with Salts and Sugars:** Often, salts and sugars are added during the canning process. These ingredients act as additional hurdles for microbial growth. Salt dehydrates bacteria by osmosis, while sugar binds to water molecules, making the environment less hospitable for microbial growth.

8. **Preserving Nutrients and Flavors:** While heat can cause some loss of nutrients, particularly water-soluble vitamins like vitamin C, canning preserves much of the nutritional value of food. It also locks in flavors, making canned products a valuable alternative to fresh, especially out of season.

Understanding how canning preserves food provides essential insight into why certain procedures and precautions must be followed during the canning process. This knowledge not only ensures safety but also enhances the quality of the preserved food, making canning a reliable and beneficial method of food preservation. As you continue to explore canning and preserving techniques, keeping these scientific principles in mind will help you achieve the best results in your home preservation efforts.

Different methods: water bath canning vs. pressure canning

Canning is a vital method for preserving food at home, and understanding the distinction between water bath canning and pressure canning is crucial for ensuring both safety and quality. These two primary canning methods cater to different types of food based on their acidity levels, and each method employs specific techniques and equipment.

Water Bath Canning

Suitable for High-Acid Foods: Water bath canning is ideal for high-acid foods, which include fruits, jams, jellies, salsas, tomatoes (with added acid), and pickles. The natural or added acids in these foods help inhibit the growth of harmful bacteria, making them safe for water bath canning.

Process:

1. **Preparation:** Prepare the food according to a tested recipe, and fill sterilized jars, leaving appropriate headspace.
2. **Sealing:** Wipe jar rims clean, apply the lid and band, and tighten to fingertip tightness.

3. **Heating:** Place jars in the canner filled with simmering water, ensuring they are completely submerged. Bring the water to a full rolling boil, then start your processing timer.
4. **Cooling:** Once processed for the recommended time, turn off the heat and let jars sit in the water for 5 minutes before removing them to cool on a towel or rack for 12 to 24 hours without disturbing.

Advantages: The equipment needed for water bath canning is relatively simple and inexpensive. It's an accessible method for beginners and is highly effective for the appropriate types of food.

Limitations: It cannot be used for low-acid foods, as the temperatures reached are not sufficient to kill the bacteria and spores that can survive in a low-acid environment.

Pressure Canning

Suitable for Low-Acid Foods: Pressure canning is necessary for low-acid foods such as meats, poultry, seafood, and most vegetables. These foods do not contain enough acid to safely inhibit bacterial growth, so they require higher temperatures than those achievable with boiling water.

Process:

1. **Preparation:** Prepare the food according to a tested recipe, and fill sterilized jars, leaving the recommended headspace.
2. **Sealing:** Wipe jar rims, apply the lid and band, and adjust to fingertip tightness.
3. **Heating:** Place jars in the pressure canner, lock the lid, and leave the vent open until steam flows freely for 10 minutes. Close the vent and bring the canner to the correct pressure for your altitude and the type of food you are canning.
4. **Cooling:** Maintain the pressure for the specific time required by the recipe. Once complete, turn off the heat and let the canner depressurize naturally. Do not open the vent. Open the canner once pressure returns to zero.

Advantages: Pressure canning enables the safe preservation of a wider variety of foods, including those that cannot be safely canned using a water bath. It ensures the destruction of all types of bacteria and spores.

Limitations: The equipment costs more and requires more careful handling to ensure safety. The process also demands more attention to detail in terms of pressure and timing.

Choosing the right canning method is dependent on the type of food you intend to preserve. Water bath canning is simpler and sufficient for fruits and acidic foods, while pressure canning is essential for preserving low-acid foods safely. Understanding these methods and their appropriate applications ensures that your home-canned foods are not only delicious but safe to consume.

The role of acidity and sugar in preservation

Acidity and sugar are crucial components in the preservation of food, particularly in canning and other preservation methods. Their role is multifaceted, affecting everything from the safety and stability of the preserved food to its flavor and texture. This section delves into how acidity and sugar contribute to the preservation process and why they are so important.

The Role of Acidity in Preservation

1. Inhibiting Microbial Growth: The pH level, which measures acidity, is critical in determining the method of canning required. High-acid foods (pH 4.6 or lower) naturally inhibit the growth of many bacteria, including the botulinum bacteria, which cannot grow in acidic environments. This makes water bath canning suitable for high-acid foods.

2. Enhancing Flavor: Acidity can enhance the natural flavors of food, making them more vibrant. It also balances the sweetness and bitterness in preserved foods.

3. Use in Pickling: Vinegar, a common acidic agent, is used extensively in pickling. The high acidity of vinegar not only preserves but also imparts the characteristic tangy flavor to pickles.

4. Safety Considerations: For foods that are on the border of high and low acidity, such as tomatoes and figs, it is often recommended to add lemon juice or vinegar to lower the pH and thus safely can them using a water bath method.

The Role of Sugar in Preservation

1. Water Activity Reduction: Sugar helps preserve food by reducing the water activity available for microbial growth. By binding to water molecules, sugar makes it difficult for microorganisms to utilize the water they need to survive.

2. Texture Preservation: Sugar helps fruits maintain their structure during the canning process, preventing them from becoming too soft. This is particularly important for high-quality preserves like jams and jellies where texture is a key attribute.

3. Flavor Enhancement: While obviously contributing sweetness, sugar also helps balance the flavors of acidic and bitter components in the food. In jams and jellies, the right amount of sugar enhances the fruit's natural flavors without overpowering them.

4. Gel Formation: In products like jellies, sugar interacts with pectin and acid to form a gel, a process that is essential for achieving the desired consistency of the final product.

Managing Acidity and Sugar in Preservation

1. Testing pH Levels: When preserving low-acid foods, it is important to ensure safety by possibly lowering the pH to safe levels using additives like citric acid or vinegar. pH strips or meters can be used to test levels after adjustments.

2. Calculating Sugar Ratios: For jams, jellies, and preserves, follow tested recipes to ensure that there is a proper balance between sugar, acid, and pectin. This balance is crucial not only for flavor and texture but also for ensuring that the preserve sets correctly.

3. Adjustments for Dietary Concerns: For those reducing sugar intake, it's possible to use sugar substitutes in some recipes, although this may affect texture and flavor. Always use recipes developed specifically for low-sugar or sugar-free preserving to ensure safety and quality.

Understanding the roles of acidity and sugar in food preservation is essential for both safety and quality. Proper management of these elements allows for the safe extension of shelf life while maintaining or enhancing the flavor and texture of the food. As you advance in your preserving skills, you'll find that mastering the balance of acidity and sugar is key to producing high-quality, safe preserved foods.

Chapter 5: Preparing Your Produce

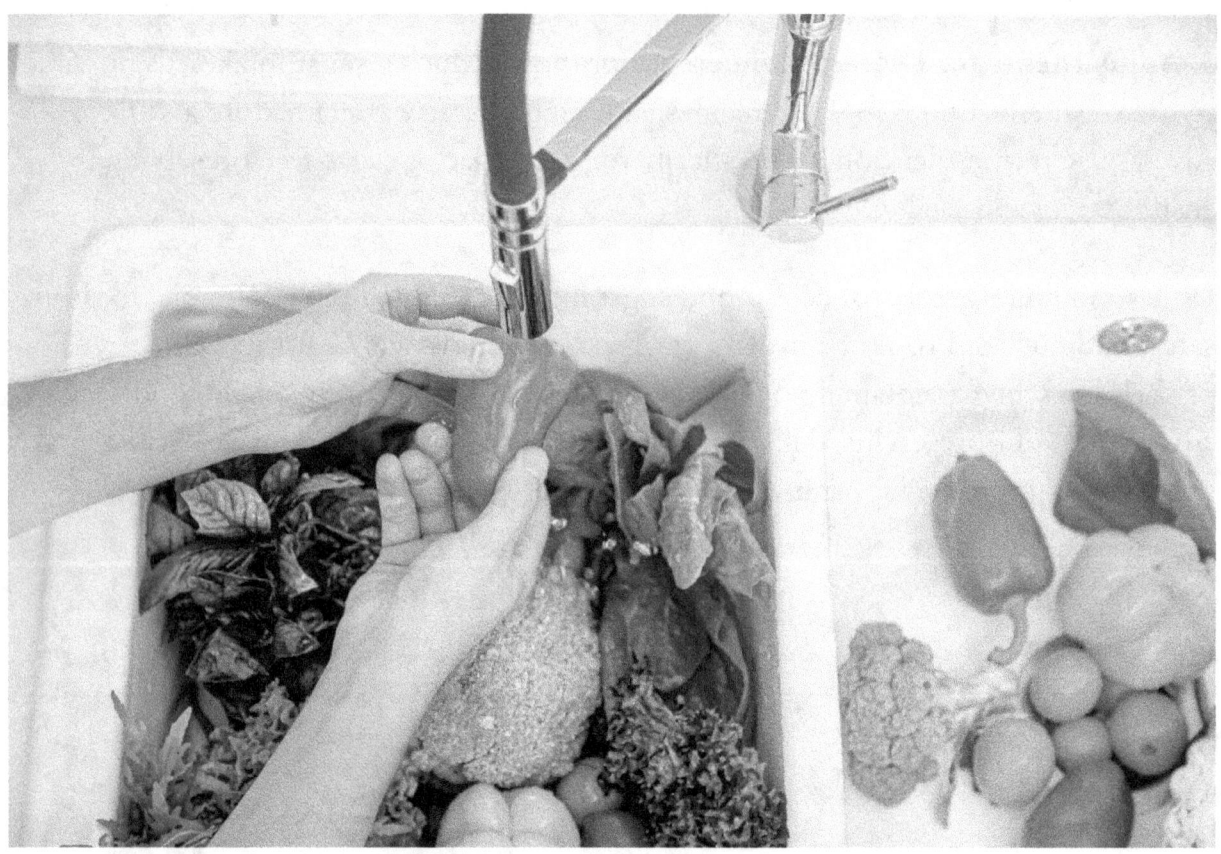

Selecting and preparing fruits and vegetables

Selecting and preparing fruits and vegetables for canning and preserving is a crucial step in the preservation process. The quality of the finished product is greatly influenced by the initial quality and preparation of the raw materials. This chapter provides detailed guidance on how to choose the best fruits and vegetables and prepare them for various preservation methods.

Selecting Fruits and Vegetables

1. Freshness: Always choose fresh, ripe, and unblemished fruits and vegetables. They should be firm, free of rot, mold, and large bruises, which can affect the flavor and safety of the preserved product.

2. Ripeness: Select fruits and vegetables at their peak ripeness. Overripe produce may be too soft and can deteriorate quickly, affecting texture and flavor, while underripe produce may lack the necessary flavor and sugar content.

3. Seasonality: Take advantage of seasonal produce when it's at its freshest and most abundant. Seasonal picking ensures the highest quality and nutritional content, as well as cost-effectiveness.

4. Organic vs. Conventional: Consider using organic produce to avoid the residues of pesticides and chemicals. If using conventional produce, ensure it is washed thoroughly.

Preparing Fruits and Vegetables

1. Washing: Rinse all fruits and vegetables under running water before canning, even if you plan to peel them. Use a soft brush to scrub items with thick skins. This step removes dirt, bacteria, and any residues on the surface.

2. Peeling and Cutting: Peel fruits and vegetables where necessary, such as peaches and tomatoes, to improve texture and appearance. Cut into uniform sizes to ensure even cooking and preservation. Removing pits, cores, and seeds is also essential.

3. Blanching: Some vegetables, like carrots and green beans, should be blanched before canning. Blanching involves boiling the vegetables for a short time and then plunging them into ice water. This process stops enzyme actions which can cause loss of flavor, color, and texture.

4. Pre-treating to Prevent Browning: Fruits like apples, pears, and peaches can brown quickly once cut. To prevent this, you can treat them with ascorbic acid (vitamin C), citric acid, or soak them in a solution of lemon juice and water.

5. Sugar Syrups and Pickling Brines: For fruits, consider preparing a light sugar syrup for canning, which helps preserve color, flavor, and texture. For vegetables to be pickled, prepare a vinegar-based brine with the appropriate concentration of vinegar to ensure safety.

Special Preparations for Specific Preserves

1. Jams and Jellies: For making jams and jellies, crush or chop the fruit into small pieces, measure accurately, and cook as directed in your recipe to achieve the perfect texture and gel.

2. Relishes and Chutneys: Chop vegetables and fruits finely and uniformly for relishes and chutneys to ensure even cooking and flavor distribution.

3. Fermentation: When preparing vegetables for fermentation, such as for sauerkraut or kimchi, cut them uniformly to ensure even fermentation. Salt is used to draw out moisture and create the brine in which the fermentation occurs.

The careful selection and preparation of fruits and vegetables are fundamental to successful canning and preserving. By starting with the best quality produce and preparing it properly, you can ensure that your preserved foods are delicious, safe, and of high quality. As you become more familiar with these processes, you'll be able to enjoy the fruits of your labor all year round.

Pre-treatment methods to enhance preservation

Pre-treatment methods play a crucial role in enhancing the quality and longevity of preserved foods. These techniques are designed to stabilize the produce, inhibit undesirable enzyme activity, and improve both the texture and flavor of the final product. This section outlines various pre-treatment methods that can be used to prepare fruits and vegetables for canning, freezing, drying, and other preservation methods.

Blanching

1. Purpose: Blanching involves briefly cooking fruits or vegetables in boiling water or steam and then rapidly cooling them in ice water. This process stops enzyme activities that can cause loss of flavor, color, and texture during storage.

2. Application: Commonly used for vegetables and some fruits before freezing. It is also used for vegetables to be canned to ensure even heat penetration during the canning process.

Syrup or Brine Preparations

1. Sugar Syrups: Immersing fruits in sugar syrup before canning can help preserve their texture, color, and flavor. The syrup can range from very light to heavy, depending on the sweetness desired and the type of fruit.

2. Pickling Brines: Vegetables intended for pickling are often soaked or cooked in a vinegar-based brine. This acidic environment is essential not just for flavor but also for inhibiting microbial growth.

Acidification

1. Purpose: Acidification is the addition of acids like lemon juice or vinegar to low-acid foods to reduce the pH to safe levels for water bath canning.

2. Application: Commonly used in canning tomatoes and figs, which may have pH values close to or above the safe threshold.

Use of Antioxidants

1. Ascorbic Acid Treatments: Ascorbic acid (vitamin C) is an effective antioxidant that prevents the browning of cut fruits. It can be used in powder form or as a solution.

2. Application: Typically used for fruits that darken quickly, such as apples, pears, and peaches, especially before freezing or drying.

Salting

1. Purpose: Salting is used to draw out moisture from food, which inhibits the growth of bacteria, yeasts, and molds. It also adds flavor.

2. Application: Often used in fermenting vegetables (e.g., sauerkraut) and in some types of pickles.

Syrup Infusion

1. Purpose: Infusing fruits in syrup at gradually increasing concentrations can help prevent shrinkage and toughening during drying.

2. Application: Ideal for fruits that are to be dried but need to retain a plump, moist texture, such as apricots and cherries.

Calcium Chloride

1. Purpose: Calcium chloride can be added to canning tomatoes or pickles to improve crispness by reinforcing the cellular structure of the fruits or vegetables.

2. Application: Used as a firming agent in canning, especially with pickles to maintain crunchiness.

These pre-treatment methods enhance the safety, quality, and sensory attributes of preserved foods. By selecting the appropriate pre-treatments based on the type of food and desired preservation method, you can significantly improve the outcome of your preservation efforts. Each method targets specific aspects of food chemistry to inhibit deterioration and maintain as much of the fresh qualities of the produce as possible. As you gain experience with these techniques, you'll be able to customize them to suit your specific needs and preferences, leading to better preserved, more enjoyable food.

Tips for maintaining flavor and texture

Preserving the flavor and texture of food during the canning and preserving process is essential to ensuring that the final product is as enjoyable as the fresh ingredients. Various techniques can be employed to maintain or even enhance these qualities. Here, we provide practical tips and strategies to help you keep the flavor and texture of your preserved foods intact.

Selecting the Right Ingredients

1. **Use Fresh and High-Quality Produce:** Always choose the freshest and highest quality fruits and vegetables. The quality of the preserved food can never exceed the quality of what you start with. Overripe or bruised produce often results in poor texture and flavor after preservation.
2. **Choose Appropriate Varieties:** Some varieties of fruits and vegetables hold up better to canning and preserving than others. For example, firm apple varieties like Granny Smith or Fuji are better for canning than softer varieties like Red Delicious.

Preparing the Ingredients

3. **Cut Uniform Pieces:** When chopping fruits and vegetables, make the pieces as uniform as possible. This not only ensures that they cook evenly but also helps maintain a consistent texture throughout the preserve.
4. **Pre-treat to Prevent Browning:** For fruits that brown easily, like apples and pears, use a pre-treatment of ascorbic acid or lemon juice. This step helps preserve the natural color and prevents the fruit from looking unappetizing.
5. **Blanch Vegetables:** Blanching vegetables before freezing or canning can help maintain vibrant colors, optimal textures, and nutritional content by halting enzyme activity that can cause spoilage.

Adjusting Cooking Times

6. **Avoid Overcooking:** Overcooking can significantly affect the texture and flavor of preserved foods. Follow the recommended cooking times precisely and use a timer to prevent this issue.
7. **Use Gentle Heat for Sensitive Items:** Foods that are delicate, such as berries and leafy greens, benefit from gentler cooking methods, such as steaming or using a water bath, to preserve their structure and taste.

Balancing Flavors

8. **Adjust Seasonings Carefully:** The flavors of spices and herbs can change intensity during the canning process. It's often better to under-season slightly

before canning and then adjust the seasoning when you open the jar to use the contents.
9. **Acidity and Sweetness:** Balance the acidity and sweetness to enhance the natural flavors of the food. For example, adding a little sugar can reduce the harshness of acidity in tomatoes, while a splash of vinegar or lemon juice can brighten overly sweet jams.

Packaging and Storage

10. **Use the Right Containers**: Make sure to use containers that are appropriate for the type of preserving you're doing. Glass jars are ideal for canning, while airtight containers or vacuum-sealed bags are best for freezing.
11. **Store Properly:** Keep preserved foods in a cool, dark place to maintain their flavor and texture. Light and heat can degrade preserved foods quickly.
12. **Monitor Storage Conditions:** Check on your stored preserves periodically for any signs of spoilage or container damage. Early detection can prevent the spread of spoilage and protect other preserved foods.

Maintaining the flavor and texture of preserved foods involves careful selection, preparation, and storage techniques. By following these tips, you can ensure that your preserved foods remain delicious and enjoyable, providing you with a taste that is as close to fresh as possible. With practice and attention to detail, you can master the art of preserving foods in ways that fully capture and retain their best qualities.

Chapter 6: Water Bath Canning

Step-by-step guide to water bath canning

Water bath canning is an accessible and effective method for preserving high-acid foods like fruits, tomatoes (with added acid), jams, jellies, salsas, and pickles. This method uses hot water to process jars of food, ensuring they are safe for storage. Here is a detailed step-by-step guide to help you successfully use the water bath canning method.

Equipment Needed

- Water bath canner or a large, deep pot with a lid
- Canning jars with new lids and bands
- Jar lifter
- Canning funnel
- Bubble remover or a non-metallic spatula

- Clean cloths or paper towels
- Timer

Preparation

1. **Prepare Your Workspace:** Clean and organize your workspace. Have all equipment and ingredients clean and ready. This includes sterilizing your jars and lids. You can sterilize jars by boiling them for 10 minutes or washing them in a dishwasher on a hot cycle with a heat dry.
2. **Prepare the Recipe:** Cook your jam, jelly, pickle, or other preserve according to your recipe. Ensure your recipe is suitable for water bath canning, particularly in terms of acidity.

Filling the Jars

3. **Fill the Jars:** Using a canning funnel, fill your hot jars with the prepared food, leaving appropriate headspace (usually about 1/4 inch to 1/2 inch, depending on the recipe). Headspace is crucial as it allows for the expansion of food during heating.
4. **Remove Air Bubbles:** Slide a non-metallic spatula or bubble remover tool around the inside edge of the jar to remove any trapped air bubbles. Adjust the headspace if necessary after removing the bubbles.
5. **Wipe the Rims:** Use a clean, damp cloth or paper towel to wipe the jar rims. Any food residue can prevent a proper seal.
6. **Apply Lids and Bands:** Place the lids on the jars, ensuring the sealing compound lines up correctly. Screw on the bands until they are fingertip tight; over-tightening can prevent air from escaping during processing, which is necessary for creating a vacuum seal.

Processing the Jars

7. **Boil Water:** Fill your canner halfway with water and preheat it. The water should be hot but not boiling when you place the jars inside.
8. **Load the Canner:** Using a jar lifter, place jars in the canner, ensuring they are not touching each other or the sides of the canner. Add more water if needed so that the jars are covered by at least 1 inch of water.

9. **Process the Jars:** Cover the pot and bring the water to a full rolling boil. Start your timer once the water is boiling. Process the jars for the time specified in your recipe, adjusting for altitude if necessary.

Cooling the Jars

10. **Remove Jars and Cool:** Once the processing time is complete, turn off the heat and remove the canner lid. Let the jars sit in the hot water for 5 minutes to adjust to temperature changes. Then, using the jar lifter, carefully lift the jars out of the water and place them on a towel or cooling rack, making sure they do not touch each other.
11. **Check Seals:** After 12 to 24 hours, check that the jars have sealed by pressing down on the center of each lid. If the lid does not pop up and down, the jar is sealed. If a jar hasn't sealed, refrigerate it and use the contents soon.
12. **Label and Store:** Label your jars with the contents and the date of canning. Store the sealed jars in a cool, dark place. Properly stored, water bath canned foods typically last up to a year.

Following these detailed steps will help ensure that your water bath canning process is successful and safe. This method is a great way to preserve the bounty of the season and enjoy your favorite fruits and vegetables year-round.

Recipes for jams, jellies, and marmalades

Jams, jellies, and marmalades are popular choices for preserving the flavors of the season in a sweet, spreadable form. Here are some classic recipes that you can try at home, each offering a unique way to enjoy fruits year-round.

1. Classic Strawberry Jam

Ingredients:

- 4 cups crushed strawberries (about 2 lbs of strawberries)

- 4 cups granulated sugar
- 1/4 cup lemon juice
- 1 packet of pectin (if using low-pectin strawberries)

Instructions:

1. Prepare your canning jars and keep them hot.
2. Combine crushed strawberries, lemon juice, and pectin in a large saucepan.
3. Bring the mixture to a full rolling boil that cannot be stirred down, over high heat, stirring constantly.
4. Add sugar, stirring to dissolve. Return the mixture to a full rolling boil. Boil hard for 1 minute, stirring constantly.
5. Remove from heat. Skim off any foam with a metal spoon.
6. Ladle immediately into prepared jars, leaving 1/4 inch headspace. Wipe rim and center lid on jar. Apply band and adjust to fingertip-tight.
7. Process in a water bath canner for 10 minutes. Remove jars and cool.

2. Orange Marmalade

Ingredients:

- 4 medium-sized oranges
- 2 lemons
- 6 cups water
- 6 cups sugar

Instructions:

1. Thinly slice the oranges and lemons, discarding the seeds. Include the peel.
2. Place the sliced fruits and water in a large pot. Let stand in a cool place for about 12 hours or overnight.
3. Bring the fruit mixture to a boil and then reduce the heat. Simmer covered for about 2 hours or until the peels are very soft.
4. Add sugar and stir until dissolved. Bring to a boil over high heat, stirring frequently. Continue to boil rapidly until the mixture reaches the gel stage (220°F on a candy thermometer).

5. Remove from heat and let stand for 5 minutes. Stir gently to distribute the peel.
6. Ladle into sterilized jars, leaving 1/4 inch headspace. Wipe the rims, apply the lids and bands, and process in a water bath canner for 10 minutes. Remove from water and cool.

3. Blackberry Jelly

Ingredients:

- 5 cups blackberry juice (about 3.5 pounds of fresh blackberries and 3/4 cup of water)
- 7 cups of sugar
- 1 pouch of liquid pectin

Instructions:

1. Extract juice by crushing blackberries and simmering them in water. Strain through a jelly bag.
2. Prepare canning jars and lids.
3. Measure juice into a large saucepot. Stir in sugar until dissolved.
4. Bring to a rolling boil over high heat. Quickly stir in liquid pectin. Return to a full rolling boil and boil hard for 1 minute, stirring constantly.
5. Remove from heat. Skim foam if necessary.
6. Quickly ladle hot jelly into prepared jars, leaving 1/4 inch headspace. Wipe rims and apply lids.
7. Process in a boiling water bath for 10 minutes. Remove jars and cool.

These recipes provide a sweet way to enjoy the fruits of your labor throughout the year. Whether you prefer the simplicity of strawberry jam, the sophisticated bitterness of orange marmalade, or the smooth texture of blackberry jelly, each recipe offers a delightful experience. Remember to follow canning guidelines carefully to ensure the safety and quality of your preserves. Enjoy spreading these homemade delights on bread, pastries, or as accompaniments to meals!

Canning high-acid foods

High-acid foods, such as most fruits, pickles, and tomatoes (with added acid), are particularly well-suited for preservation using the water bath canning method. This method utilizes the natural or added acids in foods to inhibit bacterial growth, making it a safe and effective way to preserve these types of foods. This chapter provides a detailed guide to canning high-acid foods, including key principles, safety guidelines, and best practices.

Understanding High-Acid Foods

High-acid foods are those with a pH of 4.6 or lower. The acidity can be natural, as in most fruits and some tomatoes, or added through the incorporation of vinegar or lemon juice. High acidity creates an environment that is inhospitable to the growth of most bacteria, including the botulinum bacteria.

Equipment Needed

- Water bath canner or large pot with a lid and rack
- Glass canning jars, lids, and bands
- Jar lifter
- Canning funnel
- Ladle
- Clean cloths for wiping jar rims

Step-by-Step Guide to Canning High-Acid Foods

1. Prepare Your Equipment: Sterilize your jars by boiling them in water for at least 10 minutes. Keep them hot until they are used. Lids should be washed in hot soapy water and kept warm to ensure the rubber seal is pliable.

2. Prepare the Food: Wash and prepare your high-acid food according to the recipe. This might include peeling, chopping, cooking, or mixing with other ingredients like sugar or spices.

3. Fill the Jars: Using a canning funnel, fill your hot jars with the prepared food, leaving the recommended headspace (typically 1/4 inch to 1/2 inch, depending on the food). Remove any air bubbles by gently running a non-metallic spatula around the inside edge of the jar.

4. Wipe the Rims: Clean the jar rims with a damp cloth to ensure a good seal. Any residue on the rim may prevent the lid from sealing properly.

5. Apply Lids and Bands: Place the lids on the jars, ensuring the sealing compound lines up correctly. Screw on the bands until they are fingertip tight.

6. Process in a Water Bath Canner: Place the jars in the water bath canner, making sure they are completely submerged under about 1 to 2 inches of water. Cover the canner and bring to a rolling boil. Process the jars for the time specified in your recipe, adjusting for altitude when necessary.

7. Cool the Jars: Once the processing time is complete, turn off the heat and let the jars sit in the water for 5 minutes. Remove the jars using a jar lifter and place them on a towel or cooling rack, ensuring they do not touch each other. Let them cool undisturbed for 12 to 24 hours.

8. Check Seals: After cooling, check the seals by pressing the middle of the lid. If the lid does not pop back, it is sealed. If a lid fails to seal, refrigerate the jar and use the contents within a few days.

9. Label and Store: Label your jars with the contents and the date of canning. Store in a cool, dark place. Properly canned high-acid foods typically last for up to a year.

Canning high-acid foods using the water bath method is an excellent way to preserve the bounty of the garden and enjoy seasonal flavors year-round. By following these steps, you ensure that your preserved foods are not only delicious but also safe to consume. Remember, maintaining cleanliness and adhering to processing times are critical to the safety and success of your canning efforts.

Chapter 7: Pressure Canning

Detailed instructions for pressure canning

Pressure canning is essential for preserving low-acid foods, such as vegetables, meats, poultry, and seafood. This method uses high temperature achieved by pressurized steam to destroy botulinum spores, which cannot be killed at the boiling temperature of water. Proper pressure canning is crucial to ensure the safety and quality of the food stored. Here's a step-by-step guide to pressure canning:

Equipment Needed

- Pressure canner with a locking lid, pressure gauge, and safety features
- Canning jars, new lids, and screw bands
- Jar lifter
- Canning funnel
- Bubble remover or a non-metallic spatula
- Clean cloths or paper towels

Preparation

1. **Check Your Equipment:** Before starting, inspect your pressure canner, particularly the rubber gasket, the vent, and the pressure gauge, to ensure everything is in good working condition. Replace any worn parts as needed.
2. **Sterilize Jars and Lids:** Wash jars and lids in hot soapy water and rinse well. Keep jars warm until they are ready to be used, to prevent them from breaking when hot food is added. Lids should be kept in hot (not boiling) water until used.

Filling the Jars

3. **Prepare Your Recipe:** Follow a tested pressure canning recipe. Fill your jars with the prepared food, leaving appropriate headspace as recommended in the recipe (usually between 1 to 1 1/4 inches).
4. **Remove Air Bubbles:** Run a non-metallic spatula around the inside edge of the jar to remove air bubbles. Adjust the headspace if necessary.
5. **Wipe the Rims:** Clean the jar rims with a damp cloth or paper towel to ensure no food particles interfere with the sealing process.
6. **Apply Lids and Screw Bands:** Place the lids on the jars, ensuring the sealing compound lines up correctly. Screw on the bands until fingertip tight.

Processing the Jars

7. **Prepare the Canner:** Place 2 to 3 inches of water in the bottom of the pressure canner (or as specified by the manufacturer). Place the rack in the bottom of the canner and preheat the water to 140°F for hot-packed foods or 180°F for raw-packed foods.
8. **Load the Canner:** Using a jar lifter, place filled jars on the rack in the canner. Keep the jars upright and avoid tilting them. Make sure the jars are not touching each other or the sides of the canner.
9. **Lock the Lid and Vent the Canner:** Close the canner lid securely. Leave the weight off the vent port or open the petcock. Heat at the highest setting until steam flows freely from the vent. Vent steam for 10 minutes before placing the weight on the vent or closing the petcock.

10. **Process the Jars:** Bring the canner to the correct pressure reading according to your altitude and recipe. Start timing when the correct pressure is reached. Maintain this steady pressure for the entire process time.
11. **Turn Off Heat and Cool Down:** After the processing time is complete, turn off the heat and let the canner depressurize on its own. Do not attempt to speed up the cooling process as this can result in unsafe food or jars breaking.
12. **Remove Jars:** Once the pressure has completely dropped, carefully open the canner lid away from you to avoid steam burns. Wait 10 minutes, then use a jar lifter to carefully remove the jars and place them on a towel or cooling rack, ensuring they do not touch each other.
13. **Cool and Check Seals:** Let jars cool for 12 to 24 hours. Check the seals by pressing down on the center of each lid. If the lid does not pop back, it is sealed properly. If any jars fail to seal, refrigerate them and use the contents within a few days.

Conclusion

Pressure canning is a reliable method to safely preserve low-acid foods. It requires attention to detail and strict adherence to safety guidelines to ensure that your preserved foods are safe and of high quality. By following these detailed instructions, you can confidently use your pressure canner to stock your pantry with a variety of nutritious, home-canned foods.

Safety considerations and troubleshooting

Pressure canning is a highly effective method for preserving low-acid foods, but it must be done correctly to ensure food safety. This section addresses the critical safety considerations you must adhere to while pressure canning, along with common troubleshooting tips to help you address issues that may arise during the process.

Safety Considerations

1. **Use Proper Equipment:** Only use a pressure canner that is specifically designed for canning. Pressure cookers may not maintain the correct pressure over time, which is necessary for safely canning foods.
2. **Inspect Your Equipment:** Before each use, check the pressure canner's lid, gasket, and vent pipes to ensure they are clean and free of debris. Check the rubber gasket for cracks or soft spots. Ensure that the pressure gauge is accurate by having it tested annually.
3. **Vent the Canner Properly:** Always vent the pressure canner for 10 minutes before closing the vent or placing the weight. This step is crucial to expel all the air from the canner, as the presence of air can lower the temperature achieved inside the canner, leading to unsafe canning conditions.
4. **Monitor Pressure Constantly:** Keep a close eye on the pressure gauge to ensure it stays at the correct pressure required for the foods you are canning. Fluctuating pressure can lead to improperly processed foods, which may be unsafe to eat.
5. **Adjust for Altitude:** Pressure and processing time may need to be adjusted based on your altitude. Higher altitudes require higher pressures. Refer to your canner's manual or extension services for guidelines.
6. **Use Correct Processing Times:** Follow trusted recipes and processing times. Underprocessing can lead to food spoilage and the risk of foodborne illness.
7. **Allow Natural Pressure Release:** After the processing time has elapsed, turn off the heat and allow the pressure canner to depressurize naturally. Do not try to hasten cooling by pouring cold water over the canner or opening the vent early.
8. **Handle Jars with Care:** Use a jar lifter to carefully remove jars from the canner. Avoid tilting the jars, as this can disrupt the seal.

Troubleshooting Common Issues

1. **Siphoning:** If liquid escapes from the jars during processing, it might be due to overfilling, rapid temperature changes, or fluctuating pressure. Ensure you leave adequate headspace and maintain a steady pressure.
2. **Unsealed Jars**: If jars fail to seal after cooling, check for nicks on the jar rim, ensure the lid was centered properly, and that the screw band was tightened

correctly. Reprocess the jar with a new lid within 24 hours, or refrigerate and consume the contents.

3. **Cloudy Canning Liquid:** Cloudiness can be caused by minerals in water, starch from foods, or using table salt with anti-caking agents. Use canning salt and hard water treatment if necessary.
4. **Food Spoilage:** If a jar shows signs of spoilage, such as mold, off-odors, or fermentation gases (bubbles in the jar), discard the contents safely. Always inspect the seals before consuming canned goods.
5. **Fluctuating Pressure:** If maintaining pressure is difficult, adjust the heat source gradually to stabilize the pressure and avoid rapid changes that can affect the processing.

Adhering to these safety considerations and troubleshooting tips will help ensure that your pressure canning process is both safe and effective. Remember that careful attention to detail and strict adherence to recommended guidelines are key to successful and safe home canning.

Recipes for meats, vegetables, and low-acid foods

Pressure canning is essential for safely preserving meats, vegetables, and other low-acid foods. These recipes will help you enjoy the flavors of these foods year-round and ensure they are canned safely. Each recipe is tailored for pressure canning, ensuring all safety measures are met to avoid the risk of foodborne illnesses.

1. Classic Beef Stew

Ingredients:

- 2 lbs beef stew meat, cut into 1-inch cubes
- 1 cup chopped onion
- 2 cloves garlic, minced
- 2 cups sliced carrots

- 2 cups cubed potatoes
- 1 cup sliced celery
- 1 teaspoon salt
- 1/2 teaspoon black pepper
- 1 teaspoon thyme
- Beef broth or water

Instructions:

1. Prepare pressure canner, jars, and lids.
2. Brown beef in a small amount of oil if desired.
3. Combine beef, vegetables, and spices in a large bowl. Mix well.
4. Pack the mixture into hot jars, leaving 1-inch headspace.
5. Fill jars with hot beef broth or boiling water, maintaining 1-inch headspace.
6. Remove air bubbles and adjust headspace if needed.
7. Wipe rims, apply lids, and screw on bands finger-tight.
8. Process at 10 pounds pressure (11 pounds for dial gauge canner) for 75 minutes for pints or 90 minutes for quarts, adjusting for altitude.
9. Turn off heat; let canner depressurize. Remove jars and cool.

2. Green Beans

Ingredients:

- Fresh green beans, trimmed
- Salt (optional)

Instructions:

1. Prepare pressure canner, jars, and lids.
2. Blanch green beans in boiling water for 3 minutes.
3. Pack green beans into hot jars, leaving 1-inch headspace.
4. Add 1/2 teaspoon of salt per pint, if desired.

5. Cover beans with boiling water, leaving 1-inch headspace.
6. Remove air bubbles. Wipe rims, apply lids, and screw bands on until finger-tight.
7. Process at 10 pounds pressure (11 pounds for dial gauge canner) for 20 minutes for pints or 25 minutes for quarts, adjusting for altitude.
8. Turn off heat; let canner depressurize. Remove jars and cool.

3. Chicken Soup

Ingredients:

- 2 lbs chicken, cooked and deboned
- 1 cup chopped carrots
- 1 cup chopped celery
- 1 cup diced potatoes
- 1/2 cup chopped onions
- Salt and pepper to taste
- Water or chicken broth

Instructions:

1. Prepare pressure canner, jars, and lids.
2. Combine chicken, vegetables, and seasonings in a large pot.
3. Cover with water or broth and bring to a simmer.
4. Pack hot ingredients into jars, leaving 1-inch headspace.
5. Fill jars with hot broth, leaving 1-inch headspace.
6. Remove air bubbles. Wipe rims, apply lids, and screw on bands until finger-tight.
7. Process at 10 pounds pressure (11 pounds for dial gauge canner) for 75 minutes for pints or 90 minutes for quarts, adjusting for altitude.
8. Turn off heat; let canner depressurize. Remove jars and cool.

These recipes provide a foundation for pressure canning meats, vegetables, and other low-acid foods safely. Always follow pressure canning guidelines strictly to ensure food safety. Enjoy the convenience and taste of home-canned foods throughout the year, prepared with care and preserved for enjoyment any season.

Chapter 8: Dehydration and Freezing Techniques

Basics of dehydrating fruits, vegetables, and meats

Dehydrating is a popular method for preserving food by removing moisture, which inhibits the growth of microorganisms and enzyme activity that cause spoilage. This chapter covers the basic principles of dehydrating fruits, vegetables, and meats, providing you with the foundational knowledge needed to successfully dry your own food at home.

Equipment Needed for Dehydrating

- **Dehydrator:** An electric dehydrator is the most efficient tool for drying foods, as it allows for precise temperature control and uniform air circulation.

- **Oven:** An oven can be used for dehydrating if you don't have a dehydrator. However, it is less energy-efficient and may not provide as consistent results.
- **Air-Drying Racks:** For herbs and other light items, air-drying racks in a warm, dry area can be sufficient.

Preparing Foods for Dehydration

1. Washing: Thoroughly wash all fruits, vegetables, and meats to remove any residues and microorganisms. Use clean water and, if necessary, a vegetable brush.

2. Slicing: Cut food into even, thin slices to ensure uniform drying. The thickness will depend on the type of food and the dehydrator's specifications.

3. Blanching Vegetables: Blanching vegetables before dehydrating can stop enzyme actions which can cause loss of flavor and color. Briefly boil the vegetables, then plunge them into ice water before drying.

4. Pre-treating Fruits: To prevent browning and to preserve color, pretreat fruits with ascorbic acid or lemon juice. This can be done by dipping sliced fruits into a solution of ascorbic acid and water or spraying them with lemon juice.

5. Marinating Meats for Jerky: Before dehydrating meats, marinate them to enhance flavor and to add salt, which helps in preservation. Ensure that the marinade covers the meat completely and marinate for several hours to overnight in the refrigerator.

Dehydrating Process

1. Temperature Setting: Set your dehydrator according to the type of food you are drying:

- Fruits and vegetables typically require a temperature of 125-135°F.
- Meats and fish, which need to be dried at a higher temperature to prevent bacterial growth, should be dehydrated at 145-155°F.

2. Loading the Dehydrator: Arrange the slices of food on dehydrator trays in a single layer, ensuring they do not overlap to allow for adequate air circulation.

3. Drying Time: The drying time will vary depending on the type of food, its moisture content, and the thickness of the slices:

- Fruits can take 6-36 hours.
- Vegetables can take 6-16 hours.
- Meats can take 4-12 hours.

4. Checking for Dryness: Foods should be dried until they are pliable or brittle, depending on the food. Test by cooling a piece for a few minutes before testing its texture.

Storing Dehydrated Foods

1. Conditioning: Let dried fruits "condition" for a few days before storing by placing them in a container, shaking them daily to distribute any remaining moisture evenly.

2. Packaging: Store dried foods in airtight containers, vacuum-sealed bags, or jars with tight-fitting lids to protect them from moisture and pests.

3. Cool, Dry Place: Keep the containers in a cool, dry place away from direct sunlight. Properly stored, dehydrated foods can last for several months to a year.

Conclusion

Dehydrating is a versatile preservation method that works well for a wide variety of foods and can greatly extend their shelf life. With the right preparation and drying techniques, you can enjoy the natural flavors of fruits, vegetables, and meats throughout the year. Whether you are making fruit leathers, crispy vegetable chips, or homemade jerky, dehydration offers a satisfying way to preserve your food.

How to use a dehydrator effectively

A food dehydrator is a valuable tool for preserving a wide variety of foods by removing moisture, thus inhibiting the growth of bacteria, molds, and yeast. To get the most out of your dehydrator, it's important to understand how to use it effectively. This guide provides essential tips and strategies for maximizing the efficiency of your dehydrating process.

Understanding Your Dehydrator

1. Know Your Equipment: Familiarize yourself with the specific model of dehydrator you have. Read the manufacturer's instructions carefully, as each model may have different features, capacities, and recommended temperature settings.

2. Temperature Control: Most dehydrators come with an adjustable thermostat. Understanding which temperatures are best for different types of food is crucial:

- Herbs and spices: 95°F to 105°F
- Fruits and fruit leathers: 135°F to 145°F
- Vegetables: 125°F to 135°F
- Meats and fish (jerky): 145°F to 155°F

Preparing Foods for Dehydration

3. Slice Uniformly: Ensure that all pieces of food are sliced uniformly in thickness to promote even drying. Use a mandoline slicer or a sharp knife for consistent results.

4. Pre-treatment: Some foods, like apples and pears, benefit from pre-treatment to prevent oxidation and color loss. Dip slices in lemon juice or an ascorbic acid solution before dehydrating.

5. Blanching: Blanch vegetables before dehydrating to stop enzyme activity that can cause loss of flavor and color. Blanching involves briefly boiling the vegetables and then plunging them into ice water.

Loading the Dehydrator

6. Arrange Evenly: Lay the slices of food on the dehydrator trays in a single layer without overlapping. Overlapping can cause uneven drying and increase drying time.

7. Optimize Space: While it's important not to overcrowd the trays, try to fill them as much as possible without compromising air flow to maximize energy efficiency.

During Dehydration

8. Monitoring: Check the progress periodically, especially towards the end of the drying time. Different pieces may dry at different rates based on size and water content.

9. Rotate Trays: If your dehydrator does not have a fan, rotating the trays from top to bottom and front to back can help achieve even drying.

10. Testing for Doneness: Food should be tested for dryness before being stored:

- **Fruits:** Should be pliable and leathery without any pockets of moisture.
- **Vegetables:** Should be brittle and crisp.
- **Meats:** Should be dry to the touch, pliable, and not sticky.

After Dehydration

11. Conditioning: This step is especially important for fruits to equalize the remaining moisture. Place the dried fruit in a container, loosely cover, and shake the container daily for one week.

Storage

12. Airtight Containers: Store dried foods in airtight containers in a cool, dry place. Use vacuum-sealed bags, glass jars, or plastic containers with tight-fitting lids.

13. Label and Date: Always label your containers with the type of food and the date it was dehydrated. Properly stored, dried foods can last several months to a year.

Conclusion

Using a dehydrator effectively requires understanding the nuances of how different foods dehydrate and adjusting your approach accordingly. By following these guidelines, you can ensure high-quality results and extend the shelf life of fruits, vegetables, meats, and herbs, enjoying their flavors all year round.

Freezing techniques and best practices

Freezing is one of the simplest and most effective methods for preserving food. It slows down the decomposition by turning residual moisture into ice, inhibiting the growth of most bacterial and fungal species. This chapter outlines the key techniques and best practices for freezing various types of foods effectively, ensuring they retain their quality and nutritional value.

Preparation for Freezing

1. Selecting Quality Produce: Always choose fresh, ripe, and unblemished fruits and vegetables for freezing. The quality of frozen food is only as good as the quality of the food when it was fresh.

2. Washing and Cleaning: Thoroughly wash all produce under cold running water. Avoid soaking, as this can cause some foods to lose water-soluble nutrients.

3. Blanching Vegetables: Most vegetables should be blanched before freezing. Blanching (briefly boiling and then plunging into ice water) helps to preserve color, texture, and nutritional content by stopping enzyme activity that can cause deterioration during storage.

4. Preparing Fruits: Fruits can be prepared in several ways for freezing:

- **Syrup Pack:** Ideal for soft fruits like berries, this method involves covering the fruits with a cold sugar syrup which helps preserve texture and flavor.
- **Dry Pack:** Fruits are frozen without any added sugar or liquid. This is suitable for fruits that don't clump together when frozen.
- **Sugar Pack:** Coat sliced fruits with granulated sugar; this draws out juices that create a protective syrup when frozen.

Freezing Techniques

1. Tray Packing: Spread items like berries, blanched vegetables, or individual pieces of meat on a tray to freeze them quickly. Once frozen, transfer them to airtight containers or freezer bags. This prevents the pieces from sticking together, allowing you to use portions as needed.

2. Wrapping: For meats, wrap portions tightly in freezer paper, plastic wrap, or aluminum foil, then place in freezer bags. Expel as much air as possible to prevent freezer burn.

3. Using Freezer Bags: Remove as much air as possible before sealing to minimize exposure to air, which can degrade quality and increase the risk of freezer burn.

Best Practices for Freezing

1. Labeling and Dating: Label all containers and bags with the contents and the date of freezing. Proper labeling helps manage inventory and ensures that older food is used first.

2. Maintaining Optimal Temperature: Keep your freezer at 0°F (-18°C) or below. Consistent temperatures are crucial for preserving food quality.

3. Avoid Overloading: Don't add too much unfrozen food to the freezer at one time. Overloading slows down the rate of freezing and can cause the temperature of the already frozen items to rise, which can affect their storage life.

4. Using Suitable Containers: Use containers and bags that are specifically designed for freezing. These materials are thicker and more durable, providing better protection against freezer burn and taste transfer between foods.

5. Thawing Safely: Always thaw frozen foods in the refrigerator, under cold running water, or in the microwave using the "defrost" setting. Thawing at room temperature can allow bacteria to grow.

Longevity and Quality

6. Rotation: Practice the first-in, first-out principle. Use the oldest items in your freezer first to ensure quality and reduce waste.

7. Regular Checks: Periodically check your freezer to ensure it is working correctly and that there are no signs of frost buildup or unusual ice crystals on your food, which could indicate freezer burn.

8. Avoid Refreezing: Once thawed, food should be cooked and eaten as soon as possible. Refreezing can significantly degrade the texture and flavor, and potentially allow harmful bacteria to develop.

Freezing is a versatile preservation method that works well for a wide range of foods from fruits and vegetables to meats and baked goods. By following these freezing techniques and best practices, you can maximize the lifespan and quality of your frozen foods, enjoying them as close to their original state as possible throughout the year.

Chapter 9: Fermentation and Pickling

Introduction to fermentation

Fermentation is both an art and a science, ancient in its origins yet continually evolving. It's a natural process through which microorganisms like bacteria and yeast convert organic compounds—such as sugars and starch—into alcohol or acids. This transformation not only preserves the food but also enhances its nutritional value and flavor. This chapter introduces the basics of fermentation, explaining its principles, benefits, and various applications in food preservation.

Understanding Fermentation

1. The Basics of Fermentation: At its core, fermentation is a metabolic process that produces chemical changes in organic substrates through the action of enzymes. In the context of food, it typically refers to the conversion of carbohydrates to alcohols and carbon dioxide or organic acids under anaerobic conditions (without oxygen).

2. **Types of Fermentation:**

- **Alcoholic Fermentation:** Yeasts transform sugars into alcohol and carbon dioxide. This type of fermentation is used in making beer, wine, and spirits.
- **Lactic Acid Fermentation:** Beneficial bacteria convert sugars into lactic acid, which acts as a natural preservative. This method is used in making yogurt, sauerkraut, kimchi, and sourdough bread.
- **Acetic Acid Fermentation:** Converts sugars or alcohol into acetic acid (vinegar). This is used in the production of vinegar and certain types of pickles.

Benefits of Fermentation

1. Preservation: Fermentation extends the shelf life of perishable foods by creating conditions that inhibit the growth of harmful bacteria. The acids or alcohol produced during fermentation act as natural preservatives.

2. Enhanced Nutrition: Fermenting foods can increase their nutritional value by making their nutrients more bioavailable. For example, fermentation can increase the content of beneficial enzymes, B-vitamins, omega-3 fatty acids, and various strains of probiotics.

3. Digestibility: Fermentation can break down compounds that are indigestible or cause irritation in the digestive system, such as lactose in milk. This makes fermented foods easier to digest.

4. Flavor Development: Fermentation is highly valued for the complex flavors it develops. Many of the unique flavors in foods like cheese, soy sauce, and vinegar are the result of fermentation.

Getting Started with Fermentation

1. Basic Equipment Needed:

- **Jars or Fermentation Crocks:** Clean, airtight containers are essential for maintaining the anaerobic conditions necessary for successful fermentation.
- **Weights:** To keep food submerged in its brine, which prevents mold growth by limiting exposure to air.

- **Airlocks (optional):** Devices that allow gases to escape while preventing outside air from entering.

2. **Key Ingredients:**

    - **Starter Cultures:** These can be specific strains of bacteria or yeast needed to initiate fermentation, or a brine from a previous batch.
    - **Salt:** Used in lactic acid fermentation to inhibit the growth of undesirable bacteria while promoting the growth of beneficial lactobacilli.
    - **Sugar:** Often needed to feed the yeast in alcoholic fermentation.

3. **Basic Steps:**

    - **Preparation:** Depending on the type of fermentation, this could involve mixing a brine, preparing a starter culture, or simply chopping and salting vegetables.
    - **Submersion:** Ensuring that the ingredients are fully submerged in the brine or culture solution to create an anaerobic environment.
    - **Time and Temperature:** Letting the mixture sit at room temperature, away from direct sunlight, for a period that could range from days to months, depending on the recipe.

Fermentation is a fascinating, accessible way to preserve food, enhance its nutritional profile, and develop unique flavors. As you begin to explore the various methods and recipes, you'll discover the joys of creating your own fermented foods and the benefits they bring to your health and culinary experiences. Whether it's making your own yogurt, brewing kombucha, or fermenting vegetables, the world of fermentation offers endless possibilities for creativity and flavor.

How to make pickles, sauerkraut, and other fermented foods

Fermented foods are celebrated not only for their enhanced flavors and unique textures but also for their health benefits, including improved digestion and increased nutrient absorption. This chapter provides straightforward recipes and methods for making classic fermented foods like pickles, sauerkraut, and a few others.

1. Making Classic Dill Pickles

Ingredients:

- 4 cups water
- 2 tablespoons sea salt
- 1 pound fresh cucumbers (small to medium size, preferably pickling cucumbers)
- 4-6 garlic cloves, peeled
- Fresh dill (several heads)
- 1 teaspoon mustard seeds
- 1/2 teaspoon black peppercorns
- 2-3 grape leaves (optional, to help keep pickles crisp)

Instructions:

1. **Prepare the Brine:** Dissolve sea salt in water to create a brine. Set aside.
2. **Prepare the Jars:** Sterilize the jars and lids. In each jar, place a couple of dill heads, 1-2 garlic cloves, mustard seeds, and black peppercorns.
3. **Pack the Cucumbers:** Wash the cucumbers and trim off the blossom end. Pack them into the jars tightly. Add a grape leaf on top if using.
4. **Add the Brine:** Pour the brine over the cucumbers, ensuring they are completely submerged. Leave about 1/2 inch of headspace at the top.
5. **Seal and Ferment:** Seal the jars loosely to allow gases to escape or use airlocks. Leave the jars at room temperature, away from direct sunlight, for 2-7 days. Check daily to ensure cucumbers remain submerged, and start tasting after a few days.
6. **Refrigerate:** Once the pickles reach your desired flavor and texture, tighten the lids and store them in the refrigerator. Consume within a few months.

2. Making Sauerkraut

Ingredients:

- 1 medium cabbage (about 2 pounds)
- 1 tablespoon sea salt

Instructions:

1. **Prepare the Cabbage:** Remove outer leaves of the cabbage. Slice the cabbage thinly.
2. **Salt and Massage:** Place the cabbage in a large mixing bowl, sprinkle with salt, and massage the salt into the cabbage until it starts to release liquid.
3. **Pack the Cabbage:** Transfer the cabbage into a fermentation crock or a large glass jar, a handful at a time, pressing it down firmly with your fist or a tamper to pack it tightly and force water out of it.
4. **Weigh Down the Cabbage:** Once all the cabbage is packed into the container, place a clean plate or fermentation weight on top, and press down to ensure the cabbage is submerged under its liquid.
5. **Cover and Ferment:** Cover the jar with a cloth to keep out dust and flies and secure with a rubber band. Allow the sauerkraut to ferment at room temperature for at least 2 weeks, checking it periodically to ensure it remains submerged.
6. **Store:** Once fermented, put a lid on the jar and store in the refrigerator. It will keep for several months and continue to mature and develop flavor.

3. Making Kimchi

Ingredients:

- 1 medium Napa cabbage, cut into chunks
- 1/4 cup sea salt
- Water
- 1 tablespoon grated ginger
- 4 cloves garlic, minced
- 1 teaspoon sugar

- 3 tablespoons water
- 4 tablespoons Korean red pepper flakes (gochugaru)
- 4 green onions, sliced
- 1 daikon radish, peeled and grated

Instructions:

1. **Salt the Cabbage:** Dissolve sea salt in enough water to cover the cabbage in a large bowl. Let the cabbage soak in the saltwater for about 2 hours.
2. **Make the Paste:** Blend ginger, garlic, sugar, and water into a smooth paste. Stir in gochugaru to form a spicy paste.
3. **Rinse and Mix:** Rinse the cabbage under cold water and drain. Mix the spicy paste with the cabbage, green onions, and radish.
4. **Pack and Ferment:** Pack the mixture into a clean jar, pressing down to remove air bubbles and ensure the mixture is submerged under its juice. Seal the jar and leave it at room temperature for 2-5 days for fermentation.
5. **Refrigerate and Serve:** After fermenting, store the kimchi in the refrigerator. It's ready to eat immediately but improves with age.

These recipes are just the beginning of what you can create with fermentation. Each offers a delicious way to enjoy the benefits of fermented foods and can serve as a foundation for further culinary experimentation. Whether crispy pickles, tangy sauerkraut, or spicy kimchi, these fermented foods add flavor and health benefits to any meal.

Health benefits of fermented foods

Fermented foods have been a staple in human diets around the world for thousands of years. Beyond their unique flavors and preservation benefits, these foods offer numerous health advantages. Fermentation can enhance the nutritional value of food, making it a vital part of a healthy diet. Here's an exploration of the significant health benefits associated with consuming fermented foods.

Enhanced Nutritional Content

1. Increased Vitamin Levels: Fermentation can increase the levels of certain vitamins, particularly B vitamins, including folate, riboflavin, niacin, and thiamine. Some fermented foods are also rich in vitamin K2, which is important for bone and cardiovascular health.

2. Improved Mineral Absorption: The lactic acid produced during fermentation can increase the bioavailability of minerals in foods, such as iron, zinc, and magnesium. This is because it helps to reduce levels of phytates, compounds that typically inhibit the absorption of minerals.

Digestive Health

3. Probiotics: Many fermented foods are rich in live microorganisms, known as probiotics, which can colonize the gut and help to balance the intestinal flora. This can enhance overall gut health and digestion.

4. Enzymes: Fermentation can produce digestive enzymes, which can aid in breaking down foods, making them easier to digest. This is particularly beneficial for people with digestive issues.

5. Fiber: Fermented foods can be a good source of dietary fiber, which promotes regular bowel movements and healthy digestion.

Immune System Support

6. Gut Health and Immunity: A significant portion of the immune system is located in the gut. By improving gut health, fermented foods can support a healthier immune response.

7. Anti-inflammatory Properties: Some fermented foods have anti-inflammatory properties that can help reduce chronic inflammation, a root cause of many diseases.

Weight Management

8. Weight Regulation: The improved digestion and nutrient absorption from fermented foods can help regulate appetite and body weight. Probiotics have been studied for their potential to influence fat storage and reduce obesity.

Chronic Disease Prevention

9. Cardiovascular Health: Fermented dairy products like kefir and yogurt have been linked to reduced risk of heart disease by improving lipid profiles and increasing levels of heart-healthy HDL cholesterol.

10. Type 2 Diabetes: Regular consumption of certain fermented foods can help improve insulin sensitivity and reduce the risk of type 2 diabetes.

Detoxification

11. Detoxifying Effects: Fermented foods can play a role in detoxifying the body by binding and eliminating toxins and heavy metals.

Mental Health

12. Gut-Brain Axis: Emerging research suggests that the gut microbiota has a significant impact on the brain and mental health. Consuming probiotic-rich fermented foods may positively affect mental well-being, including reducing symptoms of depression and anxiety.

The health benefits of fermented foods are vast and varied. By incorporating a variety of fermented foods into your diet, such as yogurt, kefir, sauerkraut, kimchi, and miso, you can enjoy these benefits daily. Not only do these foods enhance the flavor of your meals, but they also contribute significantly to overall health and wellness, supporting everything from digestion and immunity to mental health and chronic disease prevention.

Chapter 10: Storing and Organizing Preserved Foods

Best practices for labeling and storage

Proper labeling and storage are critical steps in the preservation process, ensuring that your efforts in canning, freezing, dehydrating, or fermenting foods are safeguarded over time. This chapter outlines best practices for labeling and storing preserved foods, helping to maintain their quality, safety, and shelf life.

Best Practices for Labeling Preserved Foods

1. Clear Labeling: Use clear, easy-to-read labels on all preserved foods. Include essential information such as the type of food, the date it was processed, and the method of preservation used (e.g., canned, frozen, dried).

2. Detailed Information: For complex recipes or items with added ingredients, list all components on the label. This can be important for tracking potential allergens or preferences when sharing or gifting.

3. Use Permanent Markers: Write with a permanent marker to ensure that labels remain legible over time. This is particularly important for items stored in environments like freezers, where moisture can cause ink to smudge.

4. Label Placement: Place labels on both the lid and the side of containers. Labeling the lid allows for easy identification when items are stacked, and side labels remain visible when containers are removed from storage.

Best Practices for Storing Preserved Foods

1. Cool, Dark, and Dry Conditions: Store preserved foods in a cool, dark, and dry place. Ideal storage temperatures are between 50°F and 70°F. Avoid areas with high temperatures or direct sunlight, which can degrade the quality of preserved foods.

2. Consistent Temperature: Fluctuating temperatures can affect the stability and safety of preserved foods, especially canned goods. Keep temperatures as constant as possible.

3. Humidity Control: Excessive humidity can lead to corrosion of metal canning lids and the growth of mold. Maintain a low-humidity environment for stored foods. Use silica gel packs or a dehumidifier in storage areas if necessary.

4. Organization: Organize preserved foods by date and type, using a "first in, first out" (FIFO) system. This ensures older items are used first, reducing waste and maintaining freshness.

5. Check Regularly for Spoilage: Regularly inspect preserved foods for signs of spoilage, such as rust on can lids, bulging lids, leaks, mold, or off odors. Discard any items that show signs of compromise.

6. Proper Freezer Management: For frozen foods, use freezer-safe containers or vacuum-sealed bags to protect against freezer burn. Organize the freezer to allow air circulation and maintain an even temperature throughout.

7. Adequate Ventilation: For items like onions and garlic stored at room temperature, ensure adequate ventilation to prevent moisture accumulation and mold growth.

8. Separate Storage Areas: Store foods with strong odors (such as spiced preserves or pickled items) away from those with milder flavors to prevent flavor transfer.

Conclusion

Adhering to these best practices for labeling and storage not only prolongs the life of your preserved foods but also helps ensure they remain safe and delicious throughout their shelf life. Proper management of your home preservation efforts will result in a reliable and enjoyable supply of home-preserved foods, ready for use in your daily cooking or emergency preparedness.

Creating an inventory system

A well-organized inventory system is essential for anyone who preserves large quantities of food. It ensures efficient use of your stock, prevents waste, and simplifies meal planning. This chapter will guide you through setting up a practical inventory system for your preserved foods, covering everything from cataloging items to tracking usage.

Designing Your Inventory System

1. Choose Your Tracking Method: Decide whether you will use a digital system, such as a spreadsheet or a dedicated app, or a manual system like a notebook. Digital systems can be more dynamic and easily searchable, while manual systems are simple and accessible without needing technology.

2. Organize by Type and Date: Group items by type (e.g., fruits, vegetables, meats) and within each type, organize by the date of preservation. This approach helps in applying the "first in, first out" (FIFO) principle, ensuring older items are used first.

3. Detailed Record-Keeping: Each entry should include the following information:

- **Type of food:** Describe the item, e.g., strawberry jam, pickled cucumbers.
- **Date of preservation:** When was the item preserved.
- **Method of preservation:** Canning, freezing, dehydrating, etc.
- **Quantity:** Number of jars, packages, or containers.
- **Expected shelf life:** How long the food is expected to remain at optimal quality.
- **Location:** Where the item is stored, especially if you use multiple storage areas or devices.

Implementing the Inventory System

4. Label and Catalogue: As you preserve new batches of food, label them clearly with all essential information and add them to your inventory list. Ensure that every member of your household knows how to read the labels and update the inventory.

5. Regular Updates: Update your inventory each time you add or remove items. Regular updates are crucial for maintaining an accurate system and avoiding surprises.

6. Periodic Reviews: Set a regular schedule to review your entire inventory. This could be monthly or seasonally, depending on how frequently you preserve food. Check for any items nearing the end of their shelf life or needing to be rotated.

7. Accessible Information: Keep your inventory records in a readily accessible place, ensuring that anyone who needs to use or update the inventory can easily do so. If using a digital system, consider backing up your data regularly.

Utilizing Your Inventory for Meal Planning

8. Plan Meals Based on Inventory: Use your inventory to plan meals, aiming to use older items or those that are plentiful. This practice helps rotate stock and reduce waste.

9. Seasonal Adjustments: Adjust your inventory system seasonally. As you enter a season of abundant fresh produce, you might freeze or can less and rely more on fresh foods. Conversely, in the off-season, your preserved foods will be a primary resource.

Benefits of an Inventory System

10. Efficient Use of Resources: An inventory helps you avoid over-purchasing or preserving more than you can realistically use, saving time and resources.

11. Enhanced Food Safety: By keeping track of the age and storage conditions of preserved foods, an inventory system helps prevent the consumption of expired or compromised food, enhancing food safety.

12. Stress Reduction: Knowing what you have on hand reduces the stress of meal planning and helps ensure you always have the ingredients you need.

Conclusion

An effective inventory system is a cornerstone of successful home preservation. It ensures that the effort and resources you invest in preserving food are maximized for benefit. By following these steps, you can create an inventory system that keeps your preserved foods organized, easily accessible, and safe to enjoy throughout the year.

Tips for maximizing shelf life

Preserving foods is a great way to extend their usability over time, but proper handling and storage are crucial to truly maximize their shelf life. Whether you're canning, freezing, dehydrating, or fermenting, following best practices can ensure your foods remain safe and tasty for as long as possible. Here are essential tips to help you maximize the shelf life of your preserved foods.

General Tips for All Preservation Methods

1. Use Fresh, High-Quality Ingredients: Always start with the freshest and highest quality produce. Spoiled or nearly spoiled foods will not preserve well and can significantly reduce the shelf life of the finished product.

2. Sterilize Equipment and Containers: Before using any canning jars, bottles, or containers, make sure they are thoroughly cleaned and sterilized to eliminate any bacteria that might cause spoilage.

3. Seal Properly: Ensure that all containers are sealed correctly to prevent air and microorganisms from entering. A proper seal is critical in preventing spoilage and extending shelf life.

4. Store in Cool, Dark Places: Light and heat can degrade preserved foods quickly, so store all items in cool, dark places like a pantry, cellar, or a dedicated storage area away from direct sunlight.

Specific Tips for Different Preservation Methods

Canning

- **Check Seals Regularly:** Inspect the seals of canned goods periodically to ensure they are still intact. Any jars that have lost their seal should be refrigerated and used quickly.
- **Avoid Freezing:** Keep canned goods in a location where they will not freeze, as freezing can cause jars to crack and seals to break.
- **Use Acidifiers:** For low-acid foods like vegetables and some meats, ensure proper acidification (e.g., adding lemon juice or vinegar) to prevent bacterial growth.

Freezing

- **Prevent Freezer Burn:** Wrap foods tightly in freezer-safe materials like heavy-duty aluminum foil, freezer paper, or plastic wrap before placing them in freezer bags. Remove as much air as possible from freezer bags to reduce the risk of freezer burn.
- **Maintain Stable Freezer Temperature:** Keep your freezer at a constant 0°F (-18°C) or below. Frequent changes in temperature can cause ice crystals to form on the food, affecting its quality.
- **Label with Freezing Date:** Always mark containers with the date of freezing to ensure you use older items first and avoid freezer burn.

Dehydrating

- **Ensure Complete Drying:** Foods should be completely dry before storage. Any residual moisture can lead to mold growth. Test by cooling a piece first; if it feels sticky or leathery, it may need more drying time.
- **Use Airtight Containers:** Store dried foods in airtight containers to keep out moisture and pests. Vacuum sealing is especially effective for long-term storage.
- **Check Periodically:** Inspect dried foods for signs of moisture or mold every few months. If you notice any condensation inside the containers, dry the contents further.

Fermenting

- **Monitor the Fermentation Process:** Keep an eye on your ferments, especially in the first few days when they are most active. Ensure that the foods are submerged in their brine, which prevents mold and yeast from forming.
- **Use Airlocks:** For long-term storage of ferments, consider using airlocks to release gases and prevent contamination from outside bacteria.
- **Refrigerate After Fermentation:** Once fermentation is complete, transfer to the refrigerator to slow down fermentation and preserve the flavor and texture of the food.

By adhering to these tips, you can significantly extend the shelf life of your preserved foods, ensuring they remain delicious and safe to consume. Proper techniques in preparation, processing, and storage are key to successful long-term preservation. Whether you're a seasoned preserver or a beginner, these practices will help you get the most out of your preservation efforts.

Chapter 11: Advanced Preserving Techniques

Advanced recipes and techniques for experienced preppers

For seasoned preppers looking to expand their repertoire and further enhance their self-sufficiency, exploring advanced recipes and techniques can be both challenging and rewarding. This chapter delves into more complex preservation methods and introduces sophisticated recipes that can help diversify your pantry and increase your preparedness for any situation.

Advanced Canning Recipes

1. Meat Stews and Soups

- **Beef Bourguignon:** Canning this classic French stew involves browning beef, mushrooms, and pearl onions, then simmering them in red wine and beef stock with herbs before canning under pressure.

- **Chicken and Vegetable Soup:** Create a hearty soup by combining chunks of chicken, diced vegetables, herbs, and chicken broth. Pressure can for long-term storage.

2. Seafood

- **Canned Salmon:** Fresh salmon can be skinned, boned, and cut into chunks. Brine, then pack it into jars with water or olive oil and pressure can.
- **Pickled Mussels:** Cook mussels in a vinegar-based pickling brine with spices, then can them for a shelf-stable, gourmet preserve.

Advanced Dehydrating Recipes

3. Fruit Leathers with Unique Flavors

- **Mango Chili Leather:** Puree ripe mangoes with lime juice and a touch of chili powder, then dehydrate the mixture into a spicy-sweet snack.
- **Berry and Mint Leather:** Blend mixed berries with fresh mint leaves and a little honey, then spread and dry for a refreshing treat.

4. Vegetable Powders

- **Tomato Powder:** Dehydrate tomato slices until crisp, then grind them into a fine powder for use in soups, stews, and sauces.
- **Garlic and Onion Powder:** Slice garlic and onions thinly, dehydrate, and grind into powder for a concentrated flavor booster.

Advanced Fermentation Techniques

5. Koji Fermentation

- **Amazake:** This sweet, fermented Japanese rice drink is made by inoculating cooked rice with koji mold. It can be consumed alone or used as a sweetener in other recipes.
- **Miso Paste:** Ferment soybeans with salt and koji to produce miso paste, a powerful flavor enhancer for soups and marinades.

6. Cheese Making

- **Aged Cheddar:** Start with pasteurized milk, add cultures and rennet, then press and age the cheese for several months to develop flavor.
- **Blue Cheese:** Incorporate Penicillium roqueforti spores into your cheese curds before pressing and aging to create the distinctive veins of blue mold.

Advanced Freezing Techniques

7. Sous Vide Cooking and Freezing

- **Sous Vide Steak:** Cook steak sous vide to the perfect doneness, quick-chill in an ice bath, then freeze. Reheat sous vide for a quick, perfectly cooked meal.
- **Freezer-Ready Meal Kits:** Assemble ingredients for meals like stir-fries or casseroles, cook using the sous vide method if appropriate, and then freeze. Thaw and reheat or finish cooking for quick, home-cooked meals.

These advanced recipes and techniques are designed for those who are already comfortable with basic preservation methods and are looking to challenge themselves and enhance their culinary options during times of need. Whether you're looking to perfect your canning skills, explore the rich flavors of fermentation, or make the most of modern conveniences like sous vide in your meal prepping, these advanced methods can significantly boost your preparedness and self-sufficiency.

Using preserves in everyday meals

Incorporating preserves into daily cooking not only enhances the flavors of everyday meals but also helps you make the most of your food storage efforts. This chapter offers creative and practical ways to use various preserves—such as jams, pickles, canned vegetables, and dried fruits—in your regular cooking routine, transforming simple dishes into rich, flavorful experiences.

Breakfast Ideas

1. Jam and Jelly:

- **Yogurt Parfait:** Layer Greek yogurt with homemade jam or jelly and a sprinkle of granola for a quick breakfast or snack.
- **Pancakes and Waffles:** Spread fruit preserves on pancakes or waffles instead of syrup for added flavor and nutrition.

2. Canned Fruit:

- **Smoothies:** Blend canned peaches or pears with a banana, some yogurt, and a dash of honey for a refreshing smoothie.
- **Oatmeal:** Stir chopped canned fruit into oatmeal during cooking; add a pinch of cinnamon for enhanced flavor.

Lunch Ideas

3. Pickles:

- **Sandwiches and Burgers:** Add pickled cucumbers, onions, or jalapeños to sandwiches and burgers for a crunchy, tangy twist.
- **Salad Dressings:** Blend pickled beets or capers into vinaigrettes for a unique salad dressing with a deep flavor profile.

4. Canned Vegetables:

- **Soup Enhancements:** Stir canned tomatoes, corn, or beans into soups and stews to bulk them up and add flavor.
- **Quick Pasta Sauce:** Sauté canned artichokes or roasted peppers with garlic, blend with cream or tomatoes, and toss with pasta for an easy meal.

Dinner Ideas

5. Canned Meat or Fish:

- **Casseroles and Pies:** Use canned chicken or tuna in pot pies or casseroles for a hearty dinner option.
- **Fish Cakes:** Mix canned salmon or sardines with breadcrumbs, eggs, and herbs, then pan-fry for delicious fish cakes.

6. Dried Fruits:

- **Tagines and Curries:** Add dried apricots or raisins to Moroccan tagines or Indian curries to introduce sweetness and texture.
- **Rice Dishes:** Stir dried cranberries or cherries into rice pilaf or couscous dishes for added color and a hint of sweetness.

Snacks and Side Dishes

7. Dehydrated Snacks:

- **Trail Mix:** Combine dehydrated apple chips, banana chips, nuts, and chocolate pieces for a custom trail mix.
- **Dips and Spreads:** Rehydrate dried tomatoes or peppers to blend into dips or spreads for crackers and bread.

8. Preserved Herbs:

- **Herb Butter:** Mix dried herbs like dill, basil, or parsley into softened butter for a flavorful spread on bread or steaks.
- **Seasoning Blends:** Grind dried herbs with salt and garlic powder to create custom seasoning blends for cooking.

Desserts

9. Jams and Fruit Preserves:

- **Fruit Tarts:** Use jams or stewed fruits as the filling in fruit tarts or turnovers.

- **Ice Cream Toppings:** Warm up jam slightly to pour over ice cream for a delicious dessert topping.

Preserves are incredibly versatile and can be used in every meal of the day, including snacks and desserts. By creatively using your stored preserves, you can enhance the taste, nutrition, and appeal of your dishes while also making efficient use of your pantry. These suggestions should inspire you to explore the wide range of possibilities that home-preserved foods offer in everyday cooking.

Making the most of seasonal abundance

Seasonal abundance provides a unique opportunity to prepare and preserve the freshest produce at its peak, allowing you to enjoy the flavors of the season throughout the year. This chapter discusses strategies for capitalizing on the bounty of each season, from identifying what to preserve, to choosing preservation methods, to incorporating these foods into your meals all year long.

Identifying Seasonal Produce

1. Know Your Seasons: Familiarize yourself with the seasonal produce in your area. **This may include:**

- **Spring:** Asparagus, strawberries, rhubarb, peas.
- **Summer:** Berries, tomatoes, cucumbers, peppers, stone fruits.
- **Fall:** Apples, pumpkins, squash, root vegetables.
- **Winter:** Citrus fruits, hardy greens, pears, nuts.

3. **Connect with Local Farmers:** Establish relationships with local farmers and frequent farmers markets to get the best selection of fresh and organic produce directly from the source.

Choosing Preservation Methods

3. Canning: Perfect for capturing the flavor of fruits and vegetables. Tomato sauces, fruit jams, and pickles are classic canned goods that preserve well.

4. Freezing: Ideal for fruits and vegetables that you want to enjoy close to their fresh state. Blanch vegetables before freezing to preserve color, texture, and nutritional value.

5. Drying: Effective for creating snacks like dried fruits, jerky, or herbs. Dehydrated foods are great for storage and can be rehydrated for cooking.

6. Fermentation: Excellent for enhancing nutritional content and creating unique flavors. Consider making sauerkraut, kimchi, or fermented pickles.

Preparing for Preservation

7. Bulk Purchasing: Take advantage of the lower prices of in-season produce by buying in bulk. This is cost-effective and allows you to preserve large quantities.

8. Immediate Processing: Preserve foods as soon as possible after harvest to maintain the highest quality and nutritional value.

9. Efficient Workflow: Organize your kitchen and workflow to handle large batches. Have all materials and equipment ready, and perhaps enlist help to manage large quantities efficiently.

Utilizing Preserved Foods

10. Cooking and Meal Planning: Integrate preserved items into your cooking throughout the year. Use fruit preserves in desserts, canned vegetables in soups and stews, and frozen items in everything from breakfasts to dinners.

11. Creative Recipes: Experiment with combining preserved ingredients in new ways. For example, add fruit compotes to savory dishes or mix different pickled vegetables for a refreshing salad.

12. Seasonal Eating: Even though you're using preserved foods, continue to cook with the seasons by integrating preserved goods with fresh produce. This approach keeps meals exciting and varied.

Maximizing Nutritional Value

13. Rotate Stock: Use older preserved items first to ensure you consume them while they are still of good quality and high nutritional value.

14. Regular Inventory Checks: Keep an inventory of what you have preserved and note when each item was stored. This helps in planning usage and ensures nothing is forgotten or wasted.

Making the most of seasonal abundance not only enhances your meals but also contributes to a sustainable lifestyle by reducing waste and dependence on out-of-season, shipped goods. By mastering the arts of preservation and creatively using your stored foods, you can enjoy the bounty of each season all year long, ensuring your meals are both delicious and nutritious.

Chapter 12: Building a Prepper's Pantry

Designing a storage space for preserved goods

Creating an effective storage space for your preserved goods is crucial for maintaining their quality and maximizing their shelf life. Whether you're storing home-canned products, dehydrated foods, or frozen preserves, a well-organized and properly designed storage area can make all the difference. Here's a guide to designing a storage space that is both functional and efficient.

Assessing Your Needs

1. Evaluate Inventory and Consumption: Before designing your storage space, assess how much you typically preserve and how quickly you consume these items. This will help you determine the amount of space needed.

2. Consider Variety: Think about the types of preservation methods you use (canning, freezing, drying, fermenting) as each type may require different storage conditions.

Choosing the Right Location

3. Cool, Dark, and Dry: For canned and dry goods, choose a space that is cool, dark, and dry. Basements, cellars, and pantries are ideal, provided they are free of dampness and not subject to extreme temperature changes.

4. Accessibility: Ensure that the storage area is easily accessible. You want to be able to reach your preserved goods without hassle, making it easier to rotate stocks and track inventory.

Designing the Layout

5. Shelving: Install sturdy shelving that can accommodate the weight of canned goods and other items. Adjustable shelves are preferable as they can be customized according to the sizes of different containers.

6. Labeling and Organization: Clearly label shelves by type of preserve or by date. Organize items in a way that follows a "first in, first out" system to use the oldest items first.

7. Maximizing Space: Use space-saving solutions like can rotators or multi-level shelves. Consider wall-mounted racks for hanging items like dried herbs or garlic braids.

Creating a Freezer Storage System

8. Dedicated Freezer Space: If freezing is a primary method of preservation for you, consider investing in a chest or upright freezer dedicated to preserved foods.

9. Organize by Type and Date: Use bins or baskets to group items by type (meat, vegetables, fruit, etc.) and label everything with contents and freezing date.

10. Avoid Overfilling: Ensure good air circulation by not overpacking the freezer. This allows for more consistent temperature control and efficient use of energy.

Special Considerations for Fermented and Sensitive Items

11. Temperature Control for Fermented Goods: Some fermented foods may need to be kept at different temperatures. If you make a lot of fermented goods, consider a temperature-controlled cabinet or a specific area in your pantry for these items.

12. Light-sensitive Items: Some preserves may be sensitive to light, such as oils or vinegars. Store these in dark bottles or containers and place them in the darkest part of your storage area.

Maintenance and Safety

13. Regular Cleaning and Maintenance: Keep your storage area clean and free of pests. Check regularly for signs of spoilage or contamination.

14. Safety First: Ensure that all electrical fittings (if any) are safely installed and that the area is free from any hazards, especially if children have access.

Designing a dedicated storage space for your preserved goods enhances the longevity and quality of your homemade products. By carefully planning the space, considering the needs of different types of preserves, and implementing an organized system, you can ensure that your pantry not only supports your preservation efforts but also contributes to an efficient and enjoyable cooking experience.

Stockpiling strategies for long-term preparedness

Stockpiling food and essential supplies is a key strategy for long-term preparedness, whether for natural disasters, economic downturns, or prolonged social disruptions.

Effective stockpiling involves more than just accumulating goods; it requires strategic planning, organization, and maintenance to ensure your supplies remain useful and accessible when needed. This chapter provides a comprehensive guide to building and managing a stockpile suited for long-term preparedness.

Assessing Needs

1. Determine Your Needs: Start by assessing how many people your stockpile will need to support and for how long. Consider special dietary requirements and preferences of each family member, including pets.

2. Calculate Consumption: Estimate the daily consumption rates of essential items such as food, water, and other necessities to project how much you'll need to store to cover a given period, typically a minimum of three months up to one year.

Building the Stockpile

3. Focus on Non-Perishables: Prioritize non-perishable food items that have a long shelf life, such as canned goods, dried beans, pasta, rice, and powdered milk. These staples form the backbone of an effective stockpile.

4. Include Ready-to-Eat Foods: Include foods that require minimal preparation, such as ready-to-eat meals, canned meats, fruits, and vegetables, which can be invaluable in situations where cooking options are limited.

5. Water Supply: Ensure you have a sufficient water supply. Store at least one gallon of water per person per day, for drinking and sanitation. Consider water purification methods and tools as part of your stockpile.

6. Diversify Your Food Sources: Incorporate a variety of nutrients into your stockpile. Include sources of protein, fats, and carbohydrates, and consider vitamin supplements to ensure balanced nutrition.

Storage Solutions

7. Choose Appropriate Storage Locations: Store your supplies in a cool, dry place out of direct sunlight. Basements, closets, and other dark, temperature-controlled areas are ideal. Avoid attics and garages if possible, as temperature fluctuations can affect the quality of stored goods.

8. Use Proper Containers: Store items in airtight containers to protect them from pests and moisture. Mylar bags with oxygen absorbers, food-grade buckets, and sealed plastic containers are excellent choices.

9. Organize for Accessibility: Organize supplies by expiration date and type. Use shelving units to keep everything accessible and ensure that older items are in front and used first to rotate stock efficiently.

Maintenance and Rotation

10. Regularly Check and Rotate Stock: Regularly inspect your stockpile for expired items, damaged containers, or signs of infestation. Rotate your supplies to use items before they expire, replacing them with newer goods.

11. Keep an Inventory: Maintain a detailed inventory of what you have, including expiration dates and quantities. This helps manage your stock more efficiently and ensures you can quickly assess your needs in an emergency.

Safety and Security

12. Secure Your Stockpile: Consider the security of your stockpile, especially in scenarios where you may be relying on these supplies for survival. Think about discreet storage solutions and security measures to protect your supplies.

13. Plan for Evacuation: In case you need to evacuate, have a plan for what to take with you. Prepare portable emergency kits and "grab-and-go" bags with essentials for each family member, including pets.

Conclusion

Effective stockpiling is a critical component of long-term preparedness, providing security and peace of mind. By carefully planning, organizing, and maintaining your stockpile, you can ensure that you and your family are well-prepared for any emergency situation. Regular reviews and updates to your plans and supplies will keep your preparedness strategy relevant and effective, no matter what challenges may arise.

Regular maintenance of the pantry

Maintaining a pantry, especially one stocked with preserved and stockpiled goods, is crucial for ensuring food safety, maximizing food quality, and managing supplies efficiently. Regular maintenance helps prevent waste, keeps your supplies fresh, and ensures you always have a clear understanding of what you have on hand. This chapter provides detailed guidelines on how to conduct regular maintenance of your pantry.

Inspection

1. Schedule Regular Checks: Set a regular schedule for inspecting your pantry. A monthly check is ideal, but at minimum, a thorough inspection should be conducted every season.

2. Check for Signs of Spoilage: Look for any signs of spoilage such as mold, off odors, leaks, or rust, especially on canned goods. Discard any items that show signs of compromise to prevent contamination of other foods.

3. Examine Packaging: Check the integrity of all packaging. Look for any tears in bags, broken seals, dents, or bulges in cans, and compromised jars (e.g., chips or cracks). Replace or repackage items as necessary to ensure long-term storage safety.

Organization

4. Rotate Stock: Use the first-in, first-out (FIFO) method to rotate your stock. This means placing newer items at the back and moving older items to the front, ensuring they are used before their expiration.

5. Organize by Category: Keep items organized by category and type. For instance, group all canned vegetables together, all spices together, etc. This not only makes items easier to find but helps in quickly assessing your stock levels in each category.

6. Accessibility: Ensure that frequently used items are easily accessible. Place less commonly used items, like seasonal or bulk goods, on higher or deeper shelves.

Inventory Management

7. Update Your Inventory List: Keep an updated inventory list that includes the quantity of items, their location, and expiration dates. This list should be reviewed and updated during each inspection.

8. Track Usage Patterns: Note which items are used most frequently and which tend to linger. This can help adjust future purchasing and preserving habits to better match your actual consumption.

Cleaning

9. Clean Shelves and Containers: During each inspection, take the time to clean shelves and storage containers. This prevents the buildup of dust and debris, which can attract pests.

10. Pest Control: Look for signs of pests such as droppings or damaged packaging. Implement measures to control pests, such as sealing cracks, using traps, or even professional pest control if necessary.

Safety Measures

11. Ensure Proper Ventilation: Good air circulation helps prevent mold and mildew. Make sure your pantry is well-ventilated and consider using a dehumidifier if moisture is an issue.

12. Check Temperature and Humidity: Regularly monitor the temperature and humidity levels in your pantry. Ideally, the pantry should be cool and dry, with temperatures between 50°F and 70°F and humidity levels below 60%.

Regular maintenance of your pantry is a vital part of household management for anyone serious about food preservation and stockpiling. By adhering to these maintenance practices, you can ensure that your pantry remains a reliable resource for high-quality, safe, and nutritious foods year-round. This proactive approach not only saves money by reducing waste but also provides peace of mind that your household is well-prepared for regular use or any emergency situations.

Conclusion

Recap of the skills and knowledge gained

Throughout this guide, we've explored a comprehensive array of topics that have equipped you with valuable skills and knowledge for effective food preservation. From the basics of canning, freezing, and dehydrating to advanced techniques in fermentation and using preserved foods in everyday meals, you've learned how to extend the shelf life of food, enhance its nutritional value, and ensure food safety. Here's a recap of the crucial skills and insights you've gained:

Fundamental Preservation Techniques

1. Canning: You've mastered both water bath and pressure canning, learning how to safely preserve high-acid foods and low-acid foods. This skill helps you store a variety of fruits, vegetables, meats, and seafood.

2. Freezing: You've learned how to properly freeze fresh produce and prepared meals, maintaining their quality and convenience. Freezing techniques, including blanching and proper packaging, have been covered to prevent freezer burn and extend shelf life.

3. Dehydrating: Techniques for drying fruits, vegetables, herbs, and meats were discussed, including how to use a dehydrator and alternative methods like oven drying. You now know how to create snacks and ingredients with concentrated flavors.

4. Fermentation: This ancient technique has been demystified, allowing you to ferment vegetables, dairy, and grains. You've learned about the health benefits of probiotics and how fermented foods can improve digestion and enhance flavors in your diet.

Advanced Food Preservation

5. Sophisticated Canning Recipes: Beyond simple jams and pickles, you've explored complex recipes like stews and seafood, expanding your canning repertoire.

6. Innovative Uses of Dehydrated Foods: We've explored how to integrate dehydrated foods into cooking and how to make unique products like fruit leathers and vegetable powders.

7. Cheese Making and Koji Fermentation: Advanced topics introduced you to the world of making your own cheeses and experimenting with koji for miso and other fermented products.

Practical Applications and Daily Use

8. Utilizing Preserves in Meals: Practical advice on incorporating preserved foods into everyday cooking ensures that you make the most out of your pantry, enhancing both the nutritional value and taste of your meals.

9. Seasonal Preserving and Stockpiling: Strategies for handling seasonal abundance and building a comprehensive stockpile have prepared you for efficient and economic food management.

Maintenance and Management

10. Storage and Organization: You've learned how to design and maintain a storage space, organize your inventory, and ensure that preserved goods remain safe and high quality over time.

11. Regular Maintenance Routines: The importance of regular checks, cleaning, and updating your inventory ensures that your pantry remains a reliable resource.

The skills and knowledge you've acquired through this guide are invaluable for increasing self-sufficiency, saving money, and enhancing your culinary options. By applying these preservation techniques, you not only ensure food security but also contribute to sustainable living practices by reducing waste and dependence on commercially processed foods. Whether you're a novice or an experienced preserver, continue to explore, experiment, and enjoy the rewards of your efforts.

Encouragement to practice and experiment

As you close this comprehensive guide on food preservation, you've armed yourself with a diverse array of techniques, from canning and freezing to dehydrating and fermenting. While the knowledge you've gained provides a solid foundation, the true mastery of these skills comes from hands-on practice and experimentation. Here's some encouragement and advice to help you continue refining your skills and exploring the boundless possibilities of food preservation.

Embrace the Learning Process

1. Start Small: If you're new to food preservation, begin with simple projects that require minimal equipment, such as making jams or freezing vegetables. Gradually build up to more complex techniques like pressure canning or fermentation.

2. Learn from Mistakes: Every mistake is a learning opportunity. Whether it's a batch of jam that didn't set or a ferment that went awry, each mishap teaches you something new about the variables and intricacies of preserving food.

Cultivate Curiosity and Creativity

3. Experiment with Flavors: Don't hesitate to tweak recipes according to your taste. Experimenting with different spices, herbs, and flavorings can transform a basic preserve into a signature concoction that could become a staple in your pantry.

4. Try New Techniques: As you become more comfortable with basic preservation methods, challenge yourself with advanced techniques like cheese making or exploring exotic fermentation practices. Each new skill you learn not only adds variety to your pantry but also broadens your culinary horizons.

Engage with a Community

5. Join Preservation Communities: Connect with others who share your interest in preservation. Online forums, local clubs, and social media groups can be invaluable resources for sharing experiences, solving problems, and finding inspiration.

6. Attend Workshops: Look for workshops and classes in your area or online. Learning from experienced instructors can provide insights that are hard to gain from books alone and can introduce you to the latest trends and techniques.

Share Your Knowledge

7. Teach Others: Sharing your skills with friends, family, or community members not only reinforces your own knowledge but also spreads the joy and benefits of food preservation. Teaching can also offer new insights as you answer questions and solve problems together.

8. Document Your Journey: Keep a journal of your preservation projects, noting what worked, what didn't, and any adjustments you made. This record can be a valuable resource as you refine your techniques.

Reflect on Your Progress

9. Celebrate Your Successes: Take time to enjoy the fruits of your labor. Whether it's tasting a perfectly balanced pickle, using your canned tomatoes in a winter stew, or sharing your dried fruit with friends, each success is a milestone worth celebrating.

10. Set New Goals: As you grow more proficient, set new goals. Perhaps you aim to preserve enough tomato sauce to last through the winter or plan to enter a local fair with your homemade preserves. New challenges keep your skills sharp and your enthusiasm high.

The art of food preservation is as rewarding as it is practical. It allows you to capture the bounty of the seasons, minimize food waste, and ensure you always have access to wholesome, homemade food. Continue to practice, experiment, and connect with others who share your passion. With each jar sealed, batch fermented, or tray dried, you are not only preserving food—you are cultivating a self-reliant lifestyle and enriching your culinary traditions. Keep exploring the vast and flavorful world of food preservation, and enjoy every moment of your journey.

Final thoughts on the role of preserving in prepping

As we conclude this comprehensive guide on preserving for preppers, it's essential to reflect on the pivotal role that preservation plays in preparedness. Whether preparing for economic instability, natural disasters, or simply striving for a sustainable way of living, the ability to preserve food efficiently is a cornerstone of self-sufficiency. Here are some final thoughts on how the art and science of food preservation enrich and empower those dedicated to preparedness.

Preservation as a Pillar of Self-Sufficiency

1. Ensuring Food Security: The primary benefit of mastering food preservation is the guarantee of a stable food supply. In times of uncertainty, a well-stocked pantry can provide peace of mind and a measure of independence from commercial food sources.

2. Economic Efficiency: By preserving seasonal produce at its peak, preppers can avoid the premium prices of off-season purchases. Additionally, preserving food helps reduce waste by ensuring that excess from harvests or bulk purchases does not spoil.

3. Nutritional Benefits: Home-preserved foods can be healthier than their store-bought counterparts, as they are free from the artificial preservatives, added sugars, and sodium that are often present in commercial products. This control over ingredients is especially crucial for maintaining a balanced diet when fresh produce is less accessible.

Preservation as a Community Anchor

4. Sharing Knowledge and Resources: Preservation often fosters community connections, as preppers share techniques, recipes, and even the fruits of their labor with neighbors, friends, and fellow enthusiasts. These connections can prove invaluable, especially in times of need.

5. Teaching and Legacy Building: Passing on the skills of canning, drying, fermenting, and freezing to the next generation ensures that these important traditions continue. These practices not only teach practical skills but also instill values of patience, resourcefulness, and stewardship of the earth.

Preservation as a Lifelong Learning Journey

6. Continuous Improvement: The world of food preservation is vast and varied. There is always something new to learn, whether a technique, recipe, or a better way to store and use preserved goods. This ongoing learning process can be incredibly fulfilling and endlessly engaging.

7. Adaptation and Innovation: As technologies evolve and new challenges arise, the methods and reasons for preserving food may also change. Staying adaptable and open to innovation will help preppers continue to refine their practices and improve their preparedness strategies.

Final Encouragement

8. Celebrate the Richness of Preserved Foods: Beyond the practical aspects, preserved foods offer a sensory and cultural richness that enhances daily meals and special occasions alike. The flavors captured in jams, pickles, and cured meats are reminders of the seasons and the care put into preparation.

9. Embrace the Prepper Spirit: At its heart, prepping is about anticipating the future and taking proactive steps to mitigate risks. Food preservation is a perfect embodiment of this spirit, blending foresight, skill, and dedication to care for oneself and others.

Conclusion

In closing, remember that each jar you seal, each batch you ferment, and each tray you dry is a step towards greater autonomy and security. Food preservation is not just a technique but a profound expression of foresight and independence. It empowers you to meet the future on your terms, with a pantry full of possibilities and a community of like-minded individuals ready to face whatever challenges may come. Keep preserving, keep preparing, and keep pushing the boundaries of what you can achieve.

Appendices

Troubleshooting guide

In the world of food preservation, encountering problems is a part of the learning curve. Whether you're canning, freezing, dehydrating, or fermenting, issues can arise that may compromise the quality or safety of your preserved goods. This chapter offers a comprehensive troubleshooting guide to help you identify and resolve common issues in various preservation methods.

Canning Troubleshooting

Problem: Jars not sealing

- **Causes:** Incomplete processing, rims not wiped clean, improper headspace, defective lids.
- **Solutions:** Ensure you're following the recommended processing time and method. Always wipe the jar rims before sealing and check that the headspace is correct. Use new, undamaged lids.

Problem: Cloudy jar contents

- **Causes:** Starch in vegetables, minerals in water, or using table salt.
- **Solutions:** Use pickling salt instead of table salt. Consider using distilled or filtered water if hard water is a problem.

Problem: Spoilage

- **Causes:** Underprocessing, contamination during handling, or fluctuating storage temperatures.
- **Solutions:** Make sure to process for the correct amount of time and at the right temperature. Always practice good hygiene when handling food and jars. Store canned goods in a cool, stable environment.

Freezing Troubleshooting

Problem: Freezer burn

- **Causes:** Air exposure, not enough wrapping, or frequent temperature changes.
- **Solutions:** Wrap food tightly with freezer-safe materials, ensuring all air is expelled from packaging. Keep the freezer temperature constant.

Problem: Ice crystals inside packages

- **Causes:** Fluctuating freezer temperatures or poor packaging.
- **Solutions:** Ensure your freezer is set to maintain a consistent 0°F (-18°C). Use quality, airtight containers or vacuum-seal bags.

Dehydrating Troubleshooting

Problem: Uneven drying

- **Causes:** Slices too thick or uneven, overcrowded trays.
- **Solutions:** Cut food into even, thin slices. Arrange slices in a single layer without overlap. Rotate trays if your dehydrator does not have a fan.

Problem: Mold growth

- **Causes:** Insufficient drying or humidity in storage.
- **Solutions:** Make sure food is completely dry before storing. Use airtight containers and keep in a cool, dry place. Consider using desiccant packets in storage containers.

Fermentation Troubleshooting

Problem: Mold or yeast growth

- **Causes:** Insufficient salinity, poor submersion under brine, or exposure to air.
- **Solutions:** Ensure your brine is salty enough to inhibit unwanted microbial growth. Keep all food submerged under the brine. Use weights and consider airlock lids for anaerobic fermentation.

Problem: Off flavors

- **Causes:** Contamination, over-fermentation, or under-fermentation.
- **Solutions:** Ensure all equipment and containers are sterilized before use. Taste test your ferment at different stages to determine when it's done to your liking.

Preservation is as much an art as it is a science. Problems can often be mitigated by careful attention to detail, cleanliness, and adherence to proven methods and recipes. If you encounter issues, use them as learning opportunities to refine your techniques and understand the processes better. With practice and patience, you'll be able to tackle these challenges effectively and continue to enjoy the benefits of homemade preserved foods.

Glossary of terms

Understanding the specific terminology used in food preservation is crucial for mastering the techniques and ensuring safety. This glossary provides definitions for common terms you'll encounter throughout the process of preserving foods.

1. Acidification: The process of adding acid (such as vinegar or lemon juice) to foods to lower the pH and inhibit the growth of harmful bacteria. Commonly used in canning low-acid foods.

2. Blanching: A cooking process that involves scalding vegetables in boiling water or steam for a short time and then plunging them into ice water. Blanching is used to halt enzyme activity that can cause loss of flavor, color, and texture.

3. Botulism: A rare but serious illness caused by a toxin produced by the bacterium Clostridium botulinum. It is associated with improperly canned or preserved foods.

4. Brine: A solution of water and salt used for pickling and fermenting foods. It creates an anaerobic environment that promotes the growth of beneficial bacteria and inhibits the growth of harmful organisms.

5. Canning: The process of preserving foods by packing them in containers and heating them to a temperature that destroys microorganisms and inactivates enzymes. The heating and subsequent cooling form a vacuum seal that prevents other microorganisms from entering and spoiling the food.

6. Dehydrating: The process of removing moisture from food by circulating hot air through a dehydrator, oven, or by sun-drying. This inhibits the growth of microorganisms and enzymes that cause spoilage.

7. Fermentation: The chemical breakdown of a substance by bacteria, yeasts, or other microorganisms, typically involving effervescence and the giving off of heat. It is used to preserve and enhance the flavor of food.

8. Freezer Burn: Dehydration and oxidation that occurs when food is not tightly sealed in the freezer. It results in dry spots and poorer quality in flavor and texture.

9. Headspace: The space left between the top of the food in a canning jar and the lid. This space is necessary to allow for the expansion of food as jars are heated and to form a vacuum seal as jars cool.

10. Jam: A type of preserve made from crushed or chopped fruits cooked with sugar until thick.

11. Jelly: A clear fruit spread made from cooked fruit juice and sugar, set by its pectin content to a firm consistency.

12. Lacto-fermentation: A method of fermentation where natural bacteria feed on the sugar and starch in the food, creating lactic acid. This process preserves the food and creates beneficial enzymes, b-vitamins, and various strains of probiotics.

13. Pectin: A substance found in the cell walls of many fruits that is used as a setting agent in jams and jellies. Pectin is extracted by cooking low-pectin fruit with high-pectin fruit or by adding commercial pectin.

14. Pickling: The process of preserving or extending the shelf life of food by either anaerobic fermentation in brine or immersion in vinegar. The resulting food is called a pickle.

15. Pressure Canning: A method of canning which uses a specialized pressure canner to achieve the high temperatures needed to safely preserve foods that are low in acid, including most vegetables and meats.

16. Water Bath Canning: A process that involves placing filled jars into a large pot of simmering water for a certain amount of time. This method is suitable for high-acid foods, including fruits, jams, jellies, and some tomatoes.

17. Vacuum Sealing: The process of removing air from a package and sealing it to extend the shelf life of foods by preventing the growth of bacteria or fungi.

By familiarizing yourself with these terms, you can enhance your understanding of food preservation and ensure that your practices are both safe and effective. This glossary serves as a valuable resource as you continue your journey in preserving foods.

THANK YOU!

collect your bonus here

scan or copy and past:

https://qrco.de/bezn6m

www.ingramcontent.com/pod-product-compliance
Lightning Source LLC
Chambersburg PA
CBHW082340220526
45470CB00008B/2579